The Development of Large Te

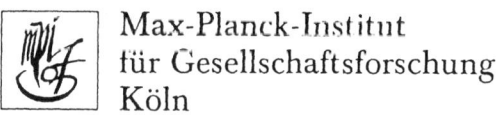

Renate Mayntz, Thomas P. Hughes (Editors)

The Development of Large Technical Systems

Campus Verlag · Frankfurt am Main
Westview Press · Boulder, Colorado

Copyright © 1988 in Frankfurt am Main by Campus Verlag

Published in 1988 in the United States by WESTVIEW PRESS
Frederick A. Praeger, Publisher
5500 Central Avenue
Boulder, Colorado 80301

ISBN 3-593-34032-1 Campus Verlag
ISBN 1-5974-0493-4 Westview Press
LCCCard Number 88-27708

Library of Congress Cataloging-in-Publication Data

The Development of large technical systems / edited by Renate Mayntz
and Thomas Hughes.
  p. cm. – (Publications of the Max Planck Institute for
Social Research, Cologne, West Germany)
  ISBN 0-8133-0839-9
  1. Technology-United States-History. 2. Technology-France-
History. 3. Technology-Germany-History. I. Mayntz, Renate.
II. Hughes, Thomas. III. Series.
T21.D49 1988
609–dc19

CIP-Titelaufnahme der Deutschen Bibliothek

The development of large technical systems / [Max-Planck-
Inst. für Gesellschaftsforschung, Köln]. Renate Mayntz;
Thomas P. Hughes (ed.). – Frankfurt am Main : Campus Verlag;
Boulder, Colorado : Westview Press, 1988
  ISBN 3-593-34032-1 (Campus) brosch.
  ISBN 0-8133-0839-9 (Westview Press) brosch.
NE: Mayntz, Renate [Hrsg.]; Max-Planck-Institut für
  Gesellschaftsforschung <Köln>

All rights reserved. No part of this book may be reproduced or transmitted in any
form or by any means, electronic or mechanical, including photocopying, recording,
or by any information storage and retrieval system, without permission in writing
from the publishers.
Printed in West Germany

CONTENTS

	Foreword	5
1.	Large technical systems: Concepts and issues *Bernward Joerges*	9
2.	The emergence of an early large-scale technical system: The American railroad network *Stephen Salsbury*	37
3.	The evolution of the technical system of railways in France from 1832 to 1937 *François Caron*	69
4.	The development of the German railroad system *G. Wolfgang Heinze and Heinrich H. Kill*	105
5.	Looking for the boundaries of technological determinism: A brief history of the U.S. telephone system *Louis Galambos*	135
6.	The telephone in France 1879 to 1979: National characteristics and international influences *Catherine Bertho-Lavenir*	155
7.	The politics of growth: The German telephone system *Frank Thomas*	179
8.	The United States air traffic system: Increasing reliability in the midst of rapid growth *Todd La Porte*	215
9.	The French electrical power system: An inter-country comparison *Maurice Lévy-Leboyer*	245
10.	The dynamics of system development in a comparative perspective: Interactive videotex in Germany, France and Britain *Renate Mayntz and Volker Schneider*	263
	Contributors	299

FOREWORD

Social science research on technology has long focused on the development, diffusion, and especially the consequences of specific isolated technologies or technical artifacts: the steam engine, the automobile, the telephone, the computer, etc. More recently, it has been recognized that an important characteristic of modern technology is the existence of complex and large technical systems - spatially extended and functionally integrated socio-technical networks such as electrical power, railroad, and telephone systems. These systems have played a focal role in the process of industrialization and economic development, and they have contributed to a significant change in life style. Aside from undoubtedly beneficial effects such systems are also creating problems - negative externalities, the risks of failure and disaster, management, control, and coordination problems. Thus a new field of research is emerging where historians and social scientists have started to cooperate in the analysis of the development and functioning of large technical systems.

The present book is the result of such cooperation. When Thomas P. Hughes published his *Networks of Power* in 1983, social scientists engaged in the study of technology reacted with immediate interest. Some were more attracted by Hughes' analysis of the social construction of technological systems, while others were more intrigued by the specific object, *large* technical *systems*, and their role in modern society. This latter interest provided the meeting-ground for Thomas P. Hughes, the historian, and Todd La Porte and Renate Mayntz, the social scientists. A joint enterprise was planned: the interdisciplinary and international study of the development, internal dynamics, management and control problems of large socio-technical systems. Since a project involving primary research on several such systems in several countries -

a necessity if theoretical generalizations are sought - did not seem feasible, the plan of a series of research conferences was developed where scholars with proven expertise on one aspect or another of this vast field of inquiry would present papers, thus collectively producing what no single researcher could have succeeded in doing.

In the summer of 1986, the Berlin Science Center hosted a small planning conference to structure the cognitive field, specify topics, and identify potential future contributors. Discussions at this planning conference lead to the identification of several sets of issues which could each serve as a topic of one research conference in the envisaged series. Taking into account the themes which the planning conference participants had formulated to indicate their own possible contributions, the development of large technical systems emerged as the best choice of a topic for the first research conference. Renate Mayntz offered to host and organize this conference. Financially supported by a grant from the German Thyssen Foundation which she obtained, the conference took place in the Max-Planck-Institut für Gesellschaftsforschung in Cologne, Germany, in November 1987. The participants were historians and social scientists specifically interested in the comparative analysis of the development of large technical systems, in particular electrical power, railroad, air traffic, telephone, and other forms of interactive telecommunication.

In contrast to other similar books which contain papers previously presented at a conference, this volume's table of contents had been planned beforehand and authors were approached to write on particular subjects, answering a set of leading questions. It was clear, moreover, that the model of systems development spelled out in *Networks of Power* would serve as a general reference point, even where no explicit comparison (as in Chapter 9) was attempted. While this does obviously not eliminate differences in analytical perspective between the sociologist, the economic historian, and the historian of technology, the similarity of approach among the authors is sufficient to warrant the claim that this is a *comparative* study of technical systems development, comparative both with respect to the technologies concerned and the national context of their implementation. This enables the reader to find answers, even if at times tentative, to such questions as

- does the development of *different* technical systems follow the same sequence of phases?

- What difference does national context (economic, legal, political) make in the development of a *given* type of technical system?
- To what extent is it possible to construct a *comprehensive* model of systems development which covers different technologies as well as different national contexts?

Renate Mayntz saw to it that the draft papers and oral presentations at the conference were transformed into the chapters that now make up this book. While she thus bears the responsibility for the final shape of this volume (and its deficiencies), it would not have been written had it not been for the work of Tom Hughes. We both thank Volker Schneider who gave valuable technical assistance in producing the print version of the book.

Renate Mayntz, Thomas P. Hughes   June 1988

CHAPTER 1
LARGE TECHNICAL SYSTEMS: CONCEPTS AND ISSUES

Bernward Joerges

> *Most persons think that a state in order to be happy ought to be large; but even if they are right, they have no idea of what is a large and what a small state... To the size of states there is a limit, as there is to other things - plants, animals, implements; for none of these retain their natural power when they are too large or too small, but they either wholly lose their nature, or are spoiled.*
> (Aristotle, in Politics)

1 Introduction

Large technical systems (LTS) are huge implements, and the public debates of the past decade or so around what is vaguely called "Big Technology" echo the age-old concern with the proper limits to the size of things. Of course, metaphors like "small is beautiful" capture many people's belief that happiness is not a matter of largeness, especially not with technical systems. But what is a large and what a small technical system? How do LTS differ from smaller techniques? Can we explain the growth and dynamics of such systems, and what does "large scale" explain? The chapter takes a broad view of conceptual issues in a social study of LTS. After shortly relating recent public controversies about "Big Technology" to the state of affairs in social science technology research, I will turn to a rare and exemplary historical approach to the study of large, integrated systems: Thomas P. Hughes's model of the evolution of local, regional and national electricity generation systems. Hughes goes far beyond mere historical description or interpretation against broad societal change. He puts forward systematic concepts generalizable to other systems of similar scale and provides a rationale for delineating *technological* systems from other social systems, small or large. This sets the scene for an examination of basic terms and explanatory issues. I will point out, with reference to current social science conceptualizations of technology, that it is far from clear what the basic terms "technical" and "large technical" mean and that for this reason the resolution of certain explanatory issues in the study of LTS meets with difficulties. Finally, LTS are exposed as a distinct type of technical system, and some conjectures as to their peculiar dynamics are offered.

It is fair to say that not only LTS, but technical phenomena in general are a neglected object of study. Social scientists have discovered public debates about "Big Technology" in the 1970s, and some have contributed to the imagery, the rhetorics and the dramatization of such debates. In the course, the term has become a *Kampfbegriff*, a battle term in the politics and "management of meaning" surrounding nuclear energy, computerization, genetic engineering and the like. In the public eye, Big Technology is high risk and high threat technology, carrying more uncertainty of consequences and more certain danger in terms of health, environmental damage, social identity, and finance - rarely in terms of political stability - than conventional production technologies. While public controversies are highly situated, and subject to marked cycles of attention, they provide the material for more comprehensive typifications and more reflexive interpretations of Big Technologies.[1] Still, the term warrants much scrutiny before it is exercised as an analytical concept.

A general question arises: Should Big Technology debates be understood as substitute debates for cultural conflict unrelated to specific LTS, or as precursor debates of more adequate sociological theorizing about them? It seems to me that public representations and rhetorics regarding BT are both: insufficient conceptually *and* substantially well directed. Consequently, social science research in this domain should aim at two things. In the first place - drawing on Durkheim - turn from "images of things" to things themselves, in this case *technical things*.[2] Secondly, specify the notion of technical *scale* and explore the implications of large and growing scale in technical and other social phenomena. This may sound trivial, but in fact implies a detour in research strategy. At present, research focuses primarily on the debates surrounding large scale technologies, just scratching the surface of LTS, as it were. In contrast, aiming at a more thorough theoretical explication of LTS as a particular type of social systems could, in time, lead back to a better understanding of the public issues they produce.

2 Networks of power

Contrary to sociology and other disciplines bent towards systematic generalization, historical approaches have produced considerable evidence for the development of specific kinds of LTS. Railroads, for instance, have been studied extensively by economic historians. The history of the emergence of large corporate organizations is inextricably linked to the large technical structures they have built up, as evidenced by Chandler's, Galambos's and Salsbury's work.[3] To the extent that such studies raise theoretical issues beyond explanation in terms of general historical forces, they tend to relate to controversies in economics or organization theory, and by the same token concern less the "technical" aspects of LTS but rather organizational structures, management strategies, economies of scale, contribution to national income and economic growth. Technology tends to be a "given".

Studies in the history of technology, by contrast, seem to have focussed very much on individual inventors and singular technical implements. It is one of Thomas P. Hughes's contributions to have brought to historical studies of technology an explicit "systems" perspective, linking technical apparatus to engineering systems, and in turn these to manifold organizational, economic and political actors and structures. Only opening up the historian-of-technology's perspective to ever larger "non-technical" contexts has allowed Hughes to embrace the complexity of evolving LTS such as, in the end, nationwide integrated electricity generation, but also other powerful networks.[4]

The starting point of Hughes's historical reconstructions, both in the sense of his initial research interest and of the elementary building block in what would later become the edifice of LTS, are successful inventor-engineers. The world of inventive engineering is seen as a peculiar world, with characteristic motive forces, resources and problem solving styles.[5] Hughes shows that the beginnings of LTS, and sometimes also their transfer, can be traced to a "type", a brand of technologists he variously calls "inventor-entrepreneurs", "independent professional inventors" or "system builders". The term "system" refers here as much to the creation, fitting together and projecting into the worlds of business, local politics and consumers of a vast number of heterogeneous technical elements as to the non-engineering activities these key actors characteristically engage in. They are as "entrepreneural" in matters technological as in dealing with outside worlds.

One may of course quarrel with Hughes that this entails a "heroization" of one group of actors - ingenious technologists - not warranted by "the facts". But his point here seems less a substantive finding than a conceptual decision. In order to understand why some attempts to install in society complex technical systems succeed while others fail, even given strong political will, business acuity, consumer demand and the like, the social nature of that subsystem he calls "technological" must be understood in the first place.

Having traced in rich detail the transition of many local to a few regional and in the end integrated electricity systems of national scope under widely varying and changing conditions in the US, France and Germany between 1880 and 1930, Hughes offers a generalized model of this process in distinguishing three main phases. The first phase goes from radical *invention*, culminating in new technological systems, through *development*, which especially involves providing technological systems with the economic and political embeddings needed for survival, to *innovation* - putting the system into efficient use. Dominant system builders in these stages are technical inventor-entrepreneurs. The next phase (which may occur at different times in the overall history of systems) is *transfer*. In order to elucidate the problems and solutions in the adaptation of systems to environments different from the ones a system has been developed in, Hughes applies the concept of "technological style": the widely varying shape "one and the same" technology takes under different geographical, political, legal and historical conditions. The concept of style also points again to the "creative latitude" of system builders, both in engineering and in organization or finance. The third phase proceeds from *growth* through *competition* to *consolidation*. Rationalization, efficiency, and capital intensification become dominant system goals, engineer-entrepreneurs are no longer in the center of activities and give way to manager-entrepreneurs and finally financier-entrepreneurs.

Again one is tempted to argue that in later phases of restructuring "mature" LTS the relative importance of "technological" protagonists may not decrease to the extent suggested by this scheme. But there is little doubt that with the growth of local systems into LTS, not only do more and more diversified actors enter into the game, but also wholly new, themselves large-scale actors such as holdings, banks, governments.

Moving up with his subject matter to ever larger systems and systems of systems, Hughes turns his conceptual focus away, then, from the shaping of technologies by identifiable actors to a great many structural features and tensions of evolving electricity systems. I will mention three: "reverse salients", "load factor", and "momentum". With the help of these concepts mainly, Hughes proposes to proceed from historical description to causal analysis, and these are the concepts that he holds useful for explaining the growth in scale of other technological systems (than electricity) as well.

As technical systems become larger, as other powerful interests and actor groups become involved in their expansion, and large organizations are built up for their gestation and drawn in from their environments, a number of phenomena and responses to them typically arise which Hughes subsumes under the term *"reverse salients"*. Reverse salients are technical or organizational anomalies resulting from uneven elaboration or evolution of a system: Progress on one front may produce backwardness elsewhere. Reverse salients require the identification and solution of underlying "critical problems" and they drive continued inventive activity and system growth. In each phase of system development, the reverse salients "elicit the emergence of a sequence of appropriate types of problem solvers, among them inventors, engineers, managers, financiers, and persons with experience in legislative and legal matters."[6] The fruitfulness of this concept lies then in its applicability to (and therefore differentiability of) dynamic processes in both technical and non-technical layers of LTS.

Hughes's application of the concept to *technical* reverse salients leads him to distinguish two types of inventions and inventors: *conservative* and *radical*. Conservative inventions, or improvements occur when critical technical problems are identified and solved by the engineering expertise of the systems's managing organizations. Radical inventions and innovations are solutions which such organizations fail to find and are instead produced by independent professional inventors. They may give rise to new, competing systems, or to the merging of separate, hitherto incompatible systems. Hughes shows that again and again it has needed "system inventors" who tended to establish themselves in independent organizations to come up with unlikely and effective solutions of reverse salients. Indeed, independent inventor-entrepreneurs could be shown to specialize in identifying critical problems and related "reverse salients" on broad technological fronts.

Load factor - the ratio of average system output to maximum output over a given period - is held, next to diversity of services and economic mix, to be a critical attribute of LTS which system builders and operators constantly try to improve. "Load factor is, probably, the major explanation for the growth of capital-intensive technological systems in capitalistic, interest-calculating societies."[7] The advantage of the concept seems to be that it refers straight to the technical core of systems, which is often masked, as Hughes says, by "concepts such as economies of scale, and motives such as drive for personal power and organizational aggrandizement."[8]

Introducing the concept of *momentum*, or dynamic inertia, Hughes leaves for good the actor perspective with which he began. Momentum seems to be a purely structural concept for capturing the unique properties that distinguish LTS from other technical systems.[9] The term aptly brings together several notions: that of giant mass, made up of innumerable technical and organizational components; of velocity, in the sense of expansiveness and rate of growth; and of goal-directedness. If reverse salients and load factor refer mainly to internal dynamics, momentum accounts for external effects. It is momentum what gives LTS the appearance of "autonomy" and deterministic power, and the concept is meant to prevent social science research on LTS to take these appearances at face value, as it were.

In his approach Hughes combines the broad frame of reference most social science disciplines would apply in studying LTS with the historian of technology's regard and respect for the technical world which the former lack. He insists that technologists and technology make a difference. And he insists that technology is far more than complicated machinery, that technologists do far more than construct machines. Bringing together two quite different sets of data - about the world-views and doings of major system inventors and about the steady if not linear emergence of giant "networks of power" - a series of conceptualizations is developed which promise to be flexible enough to accomodate other sets of data derived from other micro and macro perspectives.

Three questions may profitably be asked. Is the model compatible with evidence about these same systems produced by different approaches framed, for example, in a tradition of political economy? Can it be generalized to the generation of other LTS, such as transport systems or telecommunication systems or organ transplant systems? And can it be generalized to yet later stages in the expansion, upscaling or merging of LTS, for instance transitions to nuclear-based electricity or to satellite-based telecommunications, or to the linking of LTS into transnational systems, or the integration of separate communication networks through ISDN and the like?

As to the first point, Hughes obviously puts different questions to his material than, for instance, Perrow[10] who is interested in the way economic power structures determine technological choice. Theories are selective, and Hughes consistently applies the heuristic of tracing relationships, for example with capital, only to a point where mutual effects between capitalistic dealings and the generative mechanisms of technological systems can no longer be demonstrated in his data. But little in his model prevents linkage with economic or political models, provided these in turn conceptualize technological systems and scale as distinct phenomena.

Generalizing, secondly, to other LTS, one must keep in mind that Hughes's main body of empirical data comes from electricity in three countries. Kaijser, for instance, comparing Swedish telegraph/telephone systems with gas/electricity systems, has discovered interesting similarities and differences.[11] Both commonalities and disparities - in maturation period, justification of system integration, type of technical, capital, legal requirements and obstacles - accord well with Hughes's model. On the other hand, most studies reported in this book indicate that the phase model and the role accorded independent engineers may not fit the process dynamic of other system types, particularly in the case of implantation of new subsystems in old, "mature" LTS (see Mayntz/Schneider and their phase and structural models, this volume).

This brings up the third question, about the dynamics of LTS once they are consolidated in the sense of having reached some kind of technical monopoly (say, on a national scale), beyond Hughes's historically early stages. Before discussing some aspects of this question, however, I will take Hughes's lead of adding sociological conceptualization to biographical and historical narrative, and explore somewhat further the terminological and explanatory issues these questions present.

3 Basic terms and explanatory issues

In historical studies, the reality of technical systems is taken for granted. No need is felt to conceptualize either term, especially not the meaning of "technical", which is talked about in the language used in the texts under study. As shown, Hughes goes an unusually long way, however, in reflecting on these terms, driven by his impetus to understand the social forces that fuel the development of ever larger systems integrating technical and other parts of reality. Still, much remains implicit. Can we explicate further the terms "systems", "technical", "large" sociologically? And why should this be important?[12]

In the context of sociological studies of technology, Hughes's repeated admonition to consider companies, utilities, professional groups, financial and regulatory agencies as important components of electric power systems is paradoxically misplaced. Sociologists and political scientists, and economists too, talk only about these, happily ignoring such things as dams, turbogenerators, reactors, transformers, and large chunks of ecology. The reason is that with few exceptions materially embedded technology has not been recognized as a genuine subject matter of sociology of technology. Where technology comes in, the conceptual strategy for making it amenable to social scientific analysis is per analogy with other social phenomena.[13]

Another strategy would be "as compared with" or "as against", based on the assumption that technical systems are peculiarly different from other cultural artifacts. The task is then to spell out what makes technical systems behave differently, as opposed to organizations, ideologies, moral or legal institutions, knowledge systems, etc. LTS would, by implication, be studied as particular types of such characteristically different social entities. Explanatory issues concerning LTS would be framed in terms of more general theories of technology as phenomenon *sui generis*. The difficulty with this second strategy, which I consider superior, is the dearth of elaborated theories of technology "as against". Nevertheless, some of the arguments that follow make more sense if one adopts this stance.

"Systems"

Since most of the time this word is used in a general, nontechnical way, representing notions of complexity and complicatedness, it seems entirely unproblematic. One merit of the systems metaphor in social studies of technology is of course its endless interpretative latitude "out there": It makes sense to engineers and to disenchanted laypeople, to theoretical biologists and to grand theoreticians of society.[14] The systems term poses problems, however, when it is used more analytically. Explanatory issues where the concept promises to be useful, but begs explication, are especially connected with the micro-macro problem and the much debated problem of external versus internal control of LTS.

As to the first, one may examine questions such as: Is the systems concept useful in binding together the multi-layered phenomena identified by various partial approaches to the study of technology, in order to capture higher order interactions? How does it relate to such concepts as "interorganizational networks", *"Politikverflechtung"*, "corporatistic arrangements"? Does it help us to model linkages between complex machine systems and complex organizationnal/institutional systems? All these questions refer primarily to the inner structure of pragmatically delineated LTS.

Concerning the second issue, external versus internal control, more critical problems arise. Are LTS responding, in their build-up, to demands and requirements "outside" themselves, for example science-push, market forces, international competition, political and regulative state exigencies, even deep cultural forces? Or do LTS define scientific problems, create their markets, destroy their competitors, enlist state agencies, and shape cultural meanings? What accounts for the apparent self-propelling and potentially destructive quality, what for the counter-image of LTS as collective creations and tokens of a sustained vitality of industrial societies?

Such "locus of control" and "technological determination" issues hinge on a systems logic requiring the meaningful identification of system boundaries and substantial descriptions of system environments (or environing systems with their boundaries). The notorious ambiguity, or even emptiness, of the concepts "technical" and "technological" (systems) in sociological conceptualizations do not facilitate this. Thus Böhme, for example, rejects the systems concept in favour of what he calls "technostructure" to indicate the essential interrelatedness of linked technical objects with ecological and other social structures. Even Hughes's suggestive metaphor of the "seamless web" somehow seems to counter the systems metaphor by evoking some characteristic of never-ending generative process.[15]

"Technical systems"

It has become commonplace to say that social and technical phenomena must not be opposed, because the technical is "socially constructed", or simply because it *is* social. In practice this tends to remain lip-service, however: Enumerations such as "technological, economic, political, and social" abound. The reason is that the term "technical (as social)" remains undefined, fuzzy or residual. In other words, even as socially constructed phenomena things technical must be distinguished from socially constructed non-technical phenomena. The same applies to the distinction between "technical artifacts" and "social (artifacts)". If technical is always social, technical artifacts are social artifacts, but what kind of social artifacts are they? The legal norms governing traffic systems are social artifacts. Highways, automobiles and traffic lights are considered technical artifacts. What about the technical norms regulating road and automobile construction? Are they to be considered technical or social artifacts, or something in between?

This may sound scholastic, but it seems to me that using such labels without spelling out their relations, or collapsing them as uniformly "social", leaves us with all the pitfalls of technological or social determinisms. The situation is exacerbated by the fact that all major perspectives in present day social science theorizing about technical systems share the common feature of ignoring the material-operational cores of such systems. Explicitly or implicitly, the concept "technology" is meant to refer to phenomena other than machinery, material constructions and regulated physical processes.

In so-called contingency theories of organization, for example, "technology" figures prominently as an explanatory concept for structural features of organization, but "is not used (in the) sense of machines or sophisticated devices for achieving high efficiency..., but in its generic sense of the study of techniques or tasks."[16] But the degree of machinization of routinized task execution in, say, a traffic system is a separate concept, and measuring it independently may give rise to different interpretations. Similarly, in a social constructivist vein, systems of *human* routinized task fulfillment and *automated* systems will be invested with very different meanings; or, from a power/control perspective, capital-intensive forms of transportation pose other control (and countercontrol) problems and themselves constitute different types of resources than "labour-intensive" ones, quite irrespective of task structure.

In turn, the same machinery executing the same routinized operations, say nuclear reactors, can obviously be linked to quite diverse organizational goals and structures, will be "constructed" very differently by different user groups[17], be regulated very differently within different political regimes and represent very different economic interests[18], and so forth. Beyond showing that there is merit in combining various approaches in the study of specific technical systems, this points to the necessity of reintroducing considerations of tangible technical systems and of focussing analysis on their multiple integrations with other parts of technical and non-technical systems.

Two things would follow. In the first place, the term "social" should, if possible, be used as an umbrella term only. Non-technical social phenomena (organizations, institutions, interactions, etc.) should be specified separately. What is meant by "technical" should be specified carefully in social science terms. Otherwise its meaning will always revert to the "merely physical" aspects of technical systems.

In the second place, the freestanding material-technical artifacts and what they do by themselves should be taken seriously. This means that their actual operations (not just their design) and the actual, embodied norms (standards) governing these operations should be conceptualized as genuinely social processes of a particular kind. "Technology as knowledge" concepts have exercised the "frozen science" metaphor (Marx). But it is as important to see that technologies represent "frozen norms", institutions. Machineries are "normated natural events" (Elias), events turned into operations according to complex normative schemes (technical norms), and many recent studies of the constitution of material-technical artifacts have shown and argued that to make natural events behave is no small feat.[19]

It seems obvious to me that any specification of "technical" should be grounded in the concept of formal rationality, i.e. standardized methods of calculation on which routine actions can be based. Modern law consists of such methods, the money economy is based on such methods, and so are technical systems. While the social sciences have spent much effort in understanding legal and economic rationality and its inventions, the specific character of technical rationality has not received systematic attention. In an action theory perspective, the problem consists in establishing the *differentia specifica* of technical transactions, as opposed, for example, to legal and economic transactions. In a structural or institutional perspective, the rule systems and norms, in which such transactions are framed, would be reconstructed.

Advances in the study of technical systems rest, in this sense, on the ability (a) to conceptualize material-technical artifacts and systems of such artifacts as systems of (materialized social) action of a specific (technical) type, (b) to systematically relate such systems to other, "immaterial" systems of action (more precisely: actions mediated bodily and by talk only), (c) to differentiate, in this latter category, between systems of action which are institutionalized on the basis of formal rationalities and systems of action institutionalized on the basis of other, more inclusive cultural principles (whether these can be represented by formal models or not).

The sustainability of continued large-scale technical change will then depend, in Stinchcombe's words, on the long run "proper balance between efficient formal approximations that can have a reliable social effect, and substantive good sense to know their limits and to improve them."[20]

"Large technical systems"

Technical systems are extremely varied and everchanging, but some classification, or "social morphology" (Durkheim), is implicit in all talk about them. Large and small is a convenient basic distinction taken from public discourse and ("small is beautiful") critiques of technology. Yet, it is far from obvious what this distinction means in social science terms. Nuclear reactors are taken to be large as opposed to "small" hydropower plants, "chips" and many microelectronic implements are taken to be small and "soft". The telephone is often treated as a small-scale, everyday life technique, along with automobiles, hi-fi sets, photo cameras. Conventional weapon systems are classified as smaller than nuclear armament, and so on. But on what grounds? Even if a pragmatic approach is taken to selecting for discussion certain types of "obviously" large technical systems, this must be justifiable on more theoretical grounds. To the extent that one is interested in the specifics of LTS, that is differences between the dynamics of small-scale and large-scale technologies, the term must be derived more systematically within a broader examination of technical types.[21]

Whatever we take to be large, that which is vaguely constructed "out there" as large seems to be unsatisfactory as a delineation. Any differentiated reconstruction of the views of actors concerned with a particular system will show this. Take, for example, the telephone system as we know it. Lay users may experience telephones as quite small indeed, blissfully unaware that the telephone system might well be the largest functioning technical system yet installed in world society. Telecommunications engineers may see it as highly complex but comparatively uncomplicated, while politicians in a given country may see it as financially and legally highly complicated in view of recent technological trends, but hardly recognize its technical complexities. Future oriented corporate managers in the telecommunications industry may see it as a doomed dinosaur, social scientists just begin to appreciate the far-reaching societal impacts of past and present technical change in this field, and so on. Each of these actors, if asked to "rate" it with regard to relative scale, will produce quite different answers.

A different approach would be to "measure" scale according to the "size" of the organizations incorporating a given system. One could single out dominating agencies in a given technical system and call it large-scale if it is controlled by a large organization. Ford automobiles would be a large-scale technology, Porsche would be much smaller. Collectively, the automobile industry might come out larger than the computer or nuclear power industry.

Or, instead of referring to the powerfulness or scope of control of the central organization(s) building up and running a technical system, one could focus on "externalities", the scope of its impacts on social and natural environments. Thus, one could take up the notion prominent in public debates that "large equals risky." Or, if one feels the idea of LTS with identifiable boundaries should be given up altogether, one could turn to notions and concepts such as "seamless web", "networks of networks" and the like. Perrow's characterization of certain systems as "complex interactions plus tight coupling making for high risk potential" may be seen as a similar concept comprising both internal and external relations and providing some gauge for specifying characteristics of a particular, highly problematic subset of large technical systems.[22]

The problem with these and similar approaches to specifying what is meant by "large" seems to be the risk of confounding, in one way or another, features of large-scale technical systems which should be kept separate in order to explain their dynamics. If we declare everything *related* to a large technical system *part* of that system, including features like strong goal directedness and high growth rates, few attributes of LTS remain which could be explained by, or explain, scale. Even the devastating power of high momentum systems vis-à-vis their environments or the destructive energies set free in case of collision with similar entities are almost implied in the term - necessarily true because reflecting a purely conceptual relation[23]. Closer reference to the scale of the technical core - both materially and otherwise - of LTS would give room to inquiries about the conditions and consequences of LTS' momentum.

Risk

A growing research area devoted to the analysis of risky technical systems is mainly tuned on large systems. Two complementary explanatory strategies may be discerned, one aimed at understanding what exactly makes for system accidents[24], the other at explaining, in the first place, what makes even very large systems almost failure-free[25]. Particularly Perrow's studies of hazardous technologies have opened the field of risk analysis (occupied by engineers, cognitive psychologists, decision analysts and, to an extent, anthropologists) to more comprehensive sociological formulations and structural analysis. "System accidents", i.e. escalating failures resulting from unexpected and incalculable interactions of system components[26] are related to such structural properties as "interactive complexity" and "tight coupling". These in turn are related to intrasystem structural features such as degree of centralization or decentralization and, in an attempt to distinguish "error inducing" systems (e.g. marine transport) from "error avoiding" systems (e.g. airlines), to system environments.

In his interpretations Perrow puts much weight on power aspects, i.e. the fact that powerful organizations choose technologies which support given or desired power constellations and forms of organizational control.[27] In doing so, he stays close to the baffling public issues surrounding Big Technology. The main thrust of his argument suggests that reforming risky LTS is a political, and maybe regulative, problem, not a technical one. But much of the material presented raises another issue. Can turbulent LTS be "blackboxed" by providing them with the appropriate environment in terms of organizational and extraorganizational power structures? Or should vast and complex technical installations and the routines necessary to keep them operating be understood as social structures/institutions in their own right, no less opaque to the actors than, for example, large financial or legal structures?

Moving up to large technical infrastructures and support systems (LTS in the sense advocated here), an interesting avenue to explore would be the conceptual relationships between theories of technological risk and much more general theories of the precarious dynamics of industrial societies.[28] It is fair to say, that the fate of large financial empires and of all Western governments is very closely linked to their strategies with regard to LTS: None of them feels able to opt out of the international race to transform and radically up-scale existing LTS (energy, telecommunications, air & space), and all judge the political and financial risks of "alternative paths" as potentially disastrous - on account of their incalculability. High risk potential is almost *ex definitione* an implication of the various incalculabilities of very large (new) systems. Still it may be more useful to ask "what are the attributes of risky LTS" than "why are LTS risky". Where, then, does risk reside?

Towards a working definition

What kind of systems we eventually mean when we talk about LTS, whether it is promising to conceive of the inner structure of huge, fast and long-living material artifacts as aggregate social processes, whether we can specify and theoretically ground the notion of technical scale - answers to such questions will have to come from systematic empirical research. In the meantime, a preliminary delineation of LTS, as opposed to smaller-scale technical systems, is proposed.

I have suggested to consider technical systems as systems of machineries and freestanding structures performing, more or less reliably and predictably, complex standardized operations by virtue of being integrated with other social processes, governed and legitimated by formal, knowledge-intensive, impersonal rationalities. The guiding idea in determining the scale of such systems would then be to determine (a) the relative quantity (complexity, speed, rate of growth, etc.) of activities materialized in such systems, and (b) the quantity (complexity, speed, rate of growth) of other social processes necessitated by (a) in order to function. Of course this is not possible in any rigorous fashion.[29] Approximating a strict definition along such lines would require a lot of engineering knowledge and much preliminary and arbitrary classification of qualitatively different technical systems. However, following this notion, some types of technical systems can be singled out as undisputably large: those complex and heterogenous systems of physical structures and complex machineries which (1) are materially integrated, or "coupled" over large spans of space and time, quite irrespective of their particular cultural, political, economic and corporate make-up, and (2) support or sustain the functioning of very large numbers of other technical systems, whose organizations they thereby link.

Definitional focus on the technical core of LTS, not only in their embryonic phase but particularly in fully escalated systems, will of course interfere with arguments that "today" the old machine concept is no longer valid, that there is so much more "software" (and orgware and cultware) in technical systems now than "hardware".[30] This tends to write off the fact that there is a rather strict correspondence: the more elaborate (large) the organizational framework and cultural penetration, the more elaborate (large) the machineries and physically embodied linkages (including ecological penetration) in LTS. And, as in moving towards LTS practical engineering becomes big political negotiation and decision-making - see Hughes's system builders - so practical politics become, among other things, big engineering negotiation and decision-making.[31]

LTS in the sense suggested would be integrated transport systems, telecommunication systems, water supply systems, some energy systems, military defense systems, urban integrated public works, etc. Manufacturing technologies, single utilities, office technologies, household technologies and so forth would not qualify as LTS in this sense, no matter at which level of aggregation - except of course to the extent that they form integral parts of such systems. Arms production would not qualify, the army would; a municipal utility would not qualify, a district heating system might; personal computers in offices or even main frame systems in corporations would not qualify, a public computer network with all its components would, etc.[32]

4 A specific type of technical systems

LTS, as Hughes sees them, emerge from smaller-scale, local, intra-organizational technical systems. In growing, they undergo characteristic transformations, change from one social type into another. They also modify, merge with or supersede older LTS, which have undergone similar transmutations, gone into stasis or decline[33], or continue to compete. Going beyond models for the evolution of LTS from seminal technical systems to the dynamics of "mature", full-blown systems raises a series of additional questions.

"Organizations in technical systems"

LTS are not technical systems contained in, or co-extensive with identifiable organizations and their external reaches. Rather, LTS contain many organizations. Some of these wholly merge with LTS, others only partially, some are concerned with operating their technical subsystems, some with other, non-technical components of LTS. Other organizations just depend on their services. Those dominant actors in LTS who own, regulate or manage parts of them will be coupled (more or less loosely) politically, financially and legally. But most organizations concerned will be linked "only" technically to each other through LTS.

The latter form the social base of LTS. This base can be enormous.[34] In fact, modern LTS such as electricity systems or telecommunication systems guarantee the ongoing production, distribution, use, and disposal of almost all goods in almost all organizations of a society. They guarantee the functioning of organizations devoted to administration, health and social care, "culture", security and public order, science and education, religion and communal life. And they guarantee the functioning of all other LTS. I will not even attempt to spell out major forms of membership, access to and linkage with LTS, but suggest that this structural feature accounts for a series of other characteristics.

Retrospective studies of LTS show that they never develop according to the designs and projections of dominant actors: LTS evolve behind the backs of the system builders, as it were.[35] It has been shown, too, for instance by La Porte for national air control (this volume), that typically none of the agencies contained in LTS manage to form a somewhat complete picture of their workings. LTS seem to surpass the capacity for reflexive action of actors responsible for operating, regulating, managing and redesigning them in ways which, as social scientists, we understand poorly. How do we account for and explain the functioning and the (most of the time) relative stability of such systems?

Similarly, as long as LTS function reliably and change only incrementally, they are largely "taken for granted" by those depending on their products and services. Even partial insight into their workings is constituted mainly around failures. How is it possible then that LTS can be organized not only almost failure-free but to a high degree virtually be "silenced" and hidden away? Where blackboxing, or closure is achieved this seems to a large degree due to externalizing problems that cannot be solved within LTS to system environments, both natural and social, or by restructuring these environments. All closure is paid for by such externalities. By the same token, LTS are apt to set off far-reaching and generalized social conflicts in at least two constellations beyond the often dramatic early phases of their implantation in society. Conflict tends to be strong in cases of more or less catastrophic, and repeated, failure of major components, once this becomes to be perceived as characteristic of the entire system (in contrast, permanent unreliability in delivering services rarely seems to affect the stability of LTS). Secondly, in phases of radical reconstruction, when provisional closure becomes undone.

The polarized and at times seemingly irrational character of generalized conflict around LTS under such conditions has to do with a double tension. Taken-for-granted support systems can suddenly become very real concerns for almost everybody depending on them, and in the process the precarious nature of their closure and the cost at which it is maintained become more obvious. Some of the inner structure of large artifacts begins to show. At the same time, LTS are not "disposable" in any operational sense. Products, sites, even particular production lines and organizations can be disposed of, in the literal sense of the word, or can be redesigned and substituted. LTS as such cannot. Where generalized conflict arising from them becomes critical, closure will be sought at a new level of development. LTS seem to have an inbuilt dynamic that disposes them towards forward transformations.

LTS thus represent a societal dilemma: They are hard to blackbox for good, have an irreducible potential for controversy, and their integrations with their social base remain precarious, because for structural reasons strategies aiming at closure tend to reproduce conflict on a larger scale.

Variants of large technical systems

As a next step, it may be useful to distinguish between types of LTS and major types of subsystems. Two variants especially pose somewhat different problems, large technical networks (LTN) and large technical programs (LTP). The shuttle program, the Aswan or Cabora Bassa dams, the fusion reactor, the Chunnel project are LTP, the classical example is the Manhattan project. LTP may blend, on the one margin, into large R&D programs such as SDI, Eureca, or Esprit, or on the other into major "missing link" type projects.[36] LTN and LTP are, as a rule, closely connected and resemble each other in so far as both involve multiple state agencies, not only as regulators, and are of transnational scope; they combine, in other words, the difficulties and opportunities of multinational and mixed economy structures. Yet, the problems they pose are somewhat different.

Analyzing the linkages of participating actors to LTN, one can distinguish problems of coupling various operating and controlling actors on the one hand, and users, organizations-to-be sustained on the other hand. Control problems typically arise from a-synchronical changes on the part of diverse operating agencies, or diverse users, or *between* operators and users. The latter is of special interest because here highly variable user styles, particularly in informal everyday life settings, tend to clash with highly formalized operating and control styles. Thus, issues of load management run through the history of most LTN. Networks lack market and other regulating mechanisms allowing for efficient and flexible response. As a result, sudden or unexpected a-synchronic changes in their organizational webs "hit them harder" than, for instance, corporate production systems. Thanks mainly to their tight physical coupling, they require high central control capability and central interventions in case of failure.

Another permanent problem of LTN, akin to load problems but located at the interfaces between competing LTN or relatively autonomous subsystems, is - to generalize a term from transport planning and research - "modal split" management: the proportioning of services allocated to different (sub)systems, e.g. individual or public transport, cable or satellite transmission, human or artificial organ transplants, conventional or nuclear defense, manned or robotic space travel, and so on.

LTP, in contrast to networks, can be seen as pre-infrastructural systems oriented towards some quasi-experimental set of technical, economic or political goals. They resemble, on an over-large scale, what industrial sociologists call "stationary production processes", that is productions where design, manufacture and installation of a more or less unique industrial product are inextricably interwoven, for instance in plant construction (*Anlagenbau*).[37] In so far as LTP are often undertaken in a context of radical expansion or transformation of LTS, one may also look at them as "forward salients", if I may expand on Hughes's notion.[38] The rationale underlying LTP seems to relate one way or another to the synergetic effects to be had, or expected, from high organizational "compression". "Getting ahead", "achieving big leaps" into technical futures, often combined with strategies of pushing legal and technical standards for long-term development, are typical justifications for, as a rule, highly controversial, financially and politically risky LTP (LTP Apollo may have been an exception).

The problems of LTP arise from a need for synchronous organization and integration of "mature" and "immature" components of the end product, both with respect to hardware and "orgware". Since the products of LTP are always unicates (or reach only the prototype stage of a "normal" production cycle), design lines once embarked upon can often not be revised, for economic or, interestingly, for technical, "systems" reasons. LTP generally lack, in other words, the buffers and loops characteristic of long sequential processes in the generation, application and amplification of technical products. In this sense, one might call such programs "hyper-fast" forms of social organizations where specific sequences, from R&D through production to end use are compressed and made parallel in social time. LTP do not allow for "linearization", to adapt another term, introduced by Perrow.[39]

Once simple distinctions such as LTN and LTP prove useful, more sophisticated typologies of system structures, dynamics and problems could be envisaged. This may be done by extending conceptual frames of reference and examining the relationships of LTS within ever more inclusive spacial and temporal hierarchies (or other configurations) of higher order purpose systems, e.g. water, road, railway, air transport as members of an inclusive system devoted to the movement of people and tangible goods, itself in turn a member of a larger system comprising systems devoted to the movement of energies and information (see Heinze & Kill, this volume, for a corresponding systems logic). Or it may be done downward, by examining specific problems associated with "intrafunctional" subsystems, i.e. subsystem serving a given LTS itself, such as environmental monitoring systems or air control systems (La Porte, this volume) and with nested, "extrafunctional" subsystems serving specific system clienteles, such as SST in air traffic or videotex in the telephone system (Mayntz/Schneider, this volume).

Deep ecological penetration

All the studies in this volume bear witness to the tremendous organizational, economic, political and legal requirements of implanting LTS into society and in turn the organizational, economic etc. contraints imposed by them once they are installed. Less obvious is the lower part of the iceberg, the deep ecological penetration of LTS. To be sure, one finds numerous references to geographical and spacial factors, endowment with natural resources and similar factors in LTS studies. But again, this basis in nature lacks conceptualization, and therefore generalization. It seems useful here to apply the notion of an interpenetration of technological and ecosystems for two reasons. On the one hand, it provides a heuristic for identifying and systematizing relevant factors (which may not necessarily show up prominently in historical materials). On the other hand, it helps to remove a bias found in most historical studies of LTS, that is to focus on "natural inputs" feeding LTS build-up and to pass over their "natural outputs".

Evidently, the "ecological crisis" is almost identical with the growing burden of LTS on various ecosystems, from hydropower schemes in Brazil, water supply systems in all urban conglomerates, fossil energy generation and so on to a possible space pollution by expanding military and civil system based in outer space. The maintenance of that global LTS we call "world time" was not possible, at a certain point in time, without moving the Greenwich observatory to a less polluted site.

LTS expansion corresponds then with ever deeper ecological penetration and LTS radically affect the metabolisms of large-scale ecosystems. In turn, this has powerful ramifications, not only for the economic and political systems hosting LTS, but for the technological trajectories or "corridors" open to evolving LTS.[40]

Large-scale technical standard setting

The evolution of LTS is intimately linked to processes of technical standard setting. System construction and installation in society require all kinds of inter-industry, inter-systems and international standardization of technical norms, in order to establish and control internal and external linkages.[41] Above I have argued that technical systems should be viewed as systems of operations materialized in machineries, not only with regard to requisite knowledge, but as well to normative schemes: technical norms. Technical norms are the structural or institutional aspects of machinery: They regulate what technical artifacts are allowed to do and forced to do, and how they are allowed to interact among themselves, with people and nature. This is more obvious, for example, with respect to safety norms, meant to regulate certain impacts of machine operations on humans or the environment, or with norms directly regulating machine economy, meant to guarantee cost-effective operation. Such norms also tend to become subject to legal or quasi-legal definitions and sanctions. It is less obvious (except for engineers and machine operators) with regard to all kinds of "state-of-the-art" norms regulating the inner workings of machinery and - most important at the level of LTS - interactions among highly heterogeneous systems.

Systems of technical norms have not yet, to my knowledge, been conceptualized as a particular type of social rule systems[42]. There is a considerable body of historical studies available, though.[43] In the context of a study of LTS, complex processes of compatibilization and standardization of technical norms across organizations and machine populations remain to be explored systematically. How are large systems of machineries made to enter predictably and reliably into complex exchanges? How are relevant technical rule systems linked to other (legal, monetary) types of formalized rule systems governing LTS? Does standardization force legalization and monetarization of system relations, or is it the other way around? How do economic and legal rule systems and concepts change with the expansion of LTS in society? And what are the politics of large-scale technical standard setting, especially in transnational situations without clear "technological hegemony"? At the far end of the manifold public meanings of technology there are the out-of-awareness systems of technical norms holding together the technical cores of LTS. One will have to study the far end, too.

5 Conclusion

For quite long, the only machinery accorded conceptual status in the social sciences was the machine tool, that archetypical capital good. Recently, "the computer" has captured the theoretical imagination of sociologists of technology, mainly in its form as freestanding, self-contained machine and almost "personal" vis-à-vis. By comparison, large technical systems, in a manner of speaking the capital goods par excellence of superindustrialist society, remain undeservedly underconceptualized.

By the end of the century, many LTN retrofitted and up-scaled during the postwar decades, and many older LTS will have to be overhauled, restructured or replaced.[44] Mammoth LTP all over the globe are underway or in the offing. For better or worse, they will again change the social type of familiar, and at times almost nostalgic LTS, like the good old railroad or telephone. Computing machinery by the way will, in the form of deeply embedded control technologies, be critical in these undertakings, drive as well as constrain the up-scaling of literally all LTS.

Public perceptions and debates about their "proper size" will be studied by social scientists. But neither the perceptions nor the studies will affect the fundamental fact that modern, or if one prefers post-modern, societies irreversibly depend on the maintenance of these very large implements. Should many of them become unmanageable, on account of their catastrophic potential or because people somehow lose interest in them, they will - for simple technical reasons - in all probability not be rebuilt in human history.[45]

Notes

1 See for instance the divergent approaches illustrated by Battelle-Institut, Technologische Risiken und gesellschaftliche Konflikte. Schlussbericht: Politische Risikostrategien im Bereich der Kernenergie. Frankfurt/M., 1980, on one hand and M. Douglas and A. Wildavsky, Risk and Culture: An Essay on the Selection of Technical and Environmental Dangers. Berkeley: University of California Press, 1982, on the other.
2 This is Durkheim's advice in his first rule, see E. Durkheim, Die Regeln der soziologischen Methode. Neuwied: Luchterhand, 1895/1965, p. 115.
3 A.D. Chandler, Jr., Strategy and Structure. Cambridge: MIT Press, 1962; A.D. Chandler, Jr., The Visible Hand. Cambridge: Harvard University Press, 1977; Galambos, also Salsbury, who puts more weight on technical developments, though (this volume).
4 I refer mainly to Th.P. Hughes, Networks of Power: Electrification in Western Society 1980-1930. Baltimore/Md.: Johns Hopkins University Press, 1983; Th.P. Hughes, "The Evolution of Large Technological Systems" in W. Bijker, Th. P. Hughes and T. Pinch (eds.), The Social Construction of Technological Systems. Cambridge: M.I.T. Press, 1987, pp. 51-82.
5 See also Th.P. Hughes, "How Did the Heroic Inventors Do It?" American Heritage of Invention and Technology 1(1985)2, pp. 18-25.
6 Hughes 1987, op. cit., p. 74; see also Hughes 1983, op. cit., pp. 14-17.
7 Hughes 1987, op. cit., p. 72.
8 Ibid., p. 73.
9 At least in his latest formulation, Hughes 1987, op. cit., pp. 76-80; see however already Th.P. Hughes, "Technical Momentum in History: Hydrogenation in Germany 1898-1933", Past and Present 49(1969), pp. 106-132.
10 Ch. Perrow, Complex Organizations. 3rd ed., New York: Random House, 1986.
11 A. Kaijser, "From Local Networks to National Systems" in F. Cardot (ed.), 1880-1980: Un siècle d'électricité dans le monde. Paris: Presses Universitaires de France, 1987, pp. 7-22. As a general model of critical factors, Kaijser's "5 Ps" are offered: Price, technical Performance, Political pressures, legal Paragraphs, and public Propaganda (p. 21).
12 See R. Mayntz ("On Use and Non-use of Methodological Rules in Social Research" in U.E. Gerhard and M.E.J. Wadsworth (eds.), Stress and Stigma. Frankfurt: Campus, 1985, pp. 39-52) for a discussion of methodological issues in research on complex systems, set in the context of current controversies over analytical ("nomological") versus qualitative ("interpretive") perspectives.

13 Technology has been treated as bureaucracy, as ideology, as organization, as medium like money or law, as text or symbol, as moral institution, as knowledge system. LTS by implication would have to be treated as particularly large or powerful ideologies (organizations, media, symbols, knowledge systems, etc.). Explanatory issues would be constructed much along the lines of respective research traditions.
14 For a discussion of the heterogeneous uses of the systems term in historical studies of technology see J.M. Staudenmaier, Technology's Storytellers. Cambridge: M.I.T. Press, 1985.
15 G. Böhme, "Die Technostrukturen in der Gesellschaft" in B. Lutz (ed.), Technik und sozialer Wandel. Frankfurt: Campus, 1987, pp. 53-65; in fact, Hughes does not insist on what he terms the "outside/inside dichotomy of systems" (Hughes 1987, op. cit., p. 66), whereas most social systems theoreticians would consider the system/environment distinction fundamental. Galambos as well as Mayntz/Schneider (this volume) use the system/environment distinctions to advantage in their analyses of control processes.
16 Perrow 1986, op. cit., p. 141.
17 See e.g. Douglas and Wildavsky 1983, op. cit.; O. Keck, Policymaking in a Nuclear Program. Lexington: Lexington Books, 1981; M. Thompson, "Among the Energy Tribes", Policy Sciences 17(1984), pp. 321-333.
18 E.g. H. Kitschelt, Politik und Energie. Frankfurt: Campus, 1983.
19 See M. Callon, "Society in the Making: The Study of Technology as a Tool for Sociological Analysis" in Bijker et al. (eds.) 1987, op. cit., and other studies therein. It is one of the merits of the social constructionist school to have tangible technical artifacts brought back into the purview of sociology: ships, bicycles, telescopes, electric vehicles, and the like (see also D. MacKenzie and J. Wajcman (eds.), The Social Shaping of Technology. Milton Keynes: Open University Press, 1985). But as far as I can see, the social constructivist school have not yet consistently approached large systems of such technologies; more importantly, even this school has not addressed fully the conceptual issues involved.
20 A.L. Stinchcombe, "Reason and Rationality", Sociological Theory 4(1986), p. 151.
21 The notion of social scale is much used and little defined in general theory. Hughes's critique of the economies of scale concepts (Hughes 1987, op. cit., p. 71) is echoed in economic analyses: "Established economic theory offers virtually no significant contributions to an understanding of the past or prospective scale economies" (B. Gold, "Revising the Prevailing Approaches to Evaluating Scale Economies in Industry" in J.A. Buzacott et al. (eds.), Scale in Production Systems. Oxford: Pergamon, 1982, pp. 21-39, quoted from S.R. Carpenter's discussion "Redrawing the Bottom Line: The Optional Character of Technical Design Norms", Technology in Society 6(1984), p. 337).
22 Ch. Perrow, Normal Accidents: Living with High-Risk Technologies. New York: Basic Books, 1984.
23 R. Harré, "Philosophical aspects of the macro-micro problem" in K.J. Knorr-Cetina and A.V. Cicourel (eds.), Advances in Social Theory and Methodology, Boston: Routledge and Kegan Paul, 1981, p. 154. Empirically, the relationship between technical and organizational scale and momentum is far from clear; large-scale organizational, bureaucratic, professional and regulatory build-up and consolidation are shown to stifle system growth in many cases (see Caron, Salsbury and Heinze/Kill for railroads, Galambos for telephone, this volume).
24 Perrow 1984, op. cit.; see also Perrow 1986, op. cit., pp. 146-154.

25 T.R. La Porte et al., Studies on Complex, 'High Reliablity' Organizations. Research Note, Berkeley: University of California, 1985; La Porte, this volume.
26 Perrow's DEPOSE model: Design, Equipment, Procedures, Operations, Supplies, Environmental conditions, see Perrow 1986, op. cit., p. 147.
27 However, in my reading of Normal Accidents, he also turns up plenty of evidence for the potential of technical systems in shaping the actions of everybody concerned and creating structure which in turn prejudges technical choice, whether this is in the interest of those in power or not.
28 See especially R. Mayntz, "Die gesellschaftliche Dynamik als theoretische Herausforderung" in B. Lutz (ed.), Soziologie und gesellschaftliche Entwicklung. Frankfurt: Campus, 1985, pp. 27-44.
29 For an attempt to specify abstract properties of the concept of social scale see e.g. R.E. McGinn, "The Problem of Scale in Human Life: A Framework for Analysis", Research in Philosophy & Technology 1(1978), pp. 39-52. Regarding the technical scale of LTS, determination of their science-intensity would clearly be important (Galambos describes the paramount role of industrial research, Bertho-Lavenir of government science for systems development, this volume).
30 L. Mumford, for instance, (wishfully) talks about the "diminution of the machine" in the course of history (Technics and Civilization. New York: Harcourt, Brace & World, 1932/1963); similarly Perrow 1986, op. cit., W. Rammert, "Techniksoziologie" in G. Endruweit, B. Lutz and G. Trommersdorf (eds.), Wörterbuch der Soziologie., Stuttgart: Enke (forthcoming), 1988. A related argument in economics is that the role of "fixed capital" in industry is historically decreasing, with new flexible production technology, in favour of variable, especially "human" capital. This seems not entirely plausible.
31 See for example: L. McCartney, Friends in High Places. The Bechtel Story: The Most Secret Corporation and How it Engineered the World. New York: Simon & Schuster, 1988.
32 LTS require large-scale components (very large reactors, airports, mainframe systems, etc). In building up LTS, there seem to be lower limits to the scale of certain component systems, both technical and organizational. Large solar energy systems will require large-scale components too. Still, as with "riskiness", one should be careful to make this finding into a property ex definitione.
33 Hughes 1987, op. cit., p. 43. The decline of LTS is descibed for railroad systems by Heinze/Kill and Salsbury, in another sense by Galambos for telephony (this volume); however, this refers mainly to contracting size of systems, despite continued rapid technological change and increases in technical reach.
34 The Japanese telephone giant NTT, for instance, since its privatization by far the biggest public telecommunications corporation worldwide (if one can believe H. Becker, "Draufgänger machen mobil", Die Zeit, Nov. 6, 1987), operates 46 million terminals; it has recently found three new competitors for domestic services alone which in turn are formed by no less than 817 enterprises, among them Kyocera, Toyota Motors and the Japanese Railway Corporation.
35 In the end, it is hard to identify the "real" perpetrators of these systems, see B. Joerges, G. Bechmann and R. Hohlfeld, "Technologieentwicklung zwischen Eigendynamik und öffentlichem Diskurs: Kernenergie, Mikroelektronik und Gentechnologie in vergleichender Perspektive" in B. Lutz (ed.), Soziologie und gesellschaftliche Entwicklung. Frankfurt: Campus, 1985, pp. 355-374; J. Weyer, "European Star Wars: The Emergence of Space Technology through the Interaction of Military and Civilian Interest-Groups" in E. Mendelsohn and P. Weingart, Sociology of the Sciences, Yearbook 12, Dordrecht: Reidel, 1988 (forthcoming). Hughes recog-

nizes this in his discussion of momentum (Hughes 1987, op. cit., p. 80); see also in this volume Bertho-Lavenir for French telecom, Galambos for the beginnings of AT&T transformation, and of course the already famous case of the MINITEL...

36 Ekenger draws attention to another type of LTP: the "missing links" between national LTS, as identified for instance by the Roundtable of European Industrialists, mainly in transport and telecommunications (P. Ekenger, "Large-Scale Infrastructure Projects in Europe", Technology in Society 9(1987), pp. 87-95). This is also an instructive example for the way corporate managers "construct" LTN/LTP; see also Bertho-Lavenir's description of link-type LTP throughout the emergence of French telecom systems (this volume).

37 I owe this observation to Ingo Braun, see also the growing literature on "macro-engineering", e.g. the view from Bechtel as presented by C.W. Hull, "Macroengineering in the 1980s", Technology in Society 6(1984), pp. 45-58, where some of the largest ongoing or projected LTPs are discussed by a Bechtel CEO.

38 Hirschman, by the way, using mostly evidence from road, railroad, electricity and irrigation, has developed and elegantly formalized a general theory of such "forward and backward linkages" which seems to combine elements of both notions (A.O. Hirschman, Development Projects Observed. Washington/D.C.: The Brookings Institution, 1967).

39 Perrow 1986, op. cit., p. 147-152.

40 In this last respect, it seems probable that the technological system itself, in Hughes's sense and in its proper dynamics as presented by him, will regularly respond to ecological control issues by extending technological control, i.e. extended penetration (see also Joerges et al. 1985, op. cit., pp. 360-363).

41 Some contributions in this volume tell the story of prototypical standardization processes, e.g. Salsbury's rail gauges and automatic brakes, but especially American standard time as a "railroad invention". Hughes describes the battle between AC and DC systems in electricity (Hughes 1983, op. cit., pp. 81-90), similarly La Porte between rival radio technologies for measuring direction and distance in air traffic (this volume); for standards related battles over the integration (and exclusion) of systems see Kaijser who notes, in his study of Swedish electricity and telephony, that the transitions from local to regional to national scale "have required a thorough technical standardization of the entire systems" (From Kaijser 1987, op. cit., p. 22), Galambos (this volume) points out their role in building up the Bell system, Mayntz/Schneider (this volume) point to the hard policy games fought at the standardization level, especially in "overengineering", "techno-perfectionist" contexts such as the German, Bertho-Lavenir (this volume) highlights the political urge for technological independence in French telecommunications.

42 Openings are made in T. Burns and H. Flam, The Shaping of Social Organizations. London: Sage, 1987. In contrast, a good measure of theoretization of technical as legal norms exists, see P. Marburger, Die Regeln der Technik im Recht. Köln: Carl Heymann, 1979; also approaches in the philosophy of technology such as M. Bunge, "Technology as Applied Science", Technology and Culture 7 (1966), pp. 331-347; M. Bunge, "Philosophical Inputs and Outputs of Technology" in G. Bugliarello and D.B. Doner (eds.), The History and Philosophy of Technology, Urbana/Ill.: University of Illinois Press, 1973, pp. 262-281.

43 Recently, concepts such as Law's "metrication" (e.g. J. Law, "On the Social Explanation of Technical Change", Technology and Culture 28(1987), pp. 227-252), but also "technological testing" (E.W. Constant, "Scientific Theory and

Technological Testability", Technology and Culture 24(1983), pp. 183-198) point at genetic aspects of technical normalization.

44 See for instance Prognos AG, "Die Bundesrepublik Deutschland 1990/2000/2010", Prognos Report 12. Stuttgart: Poller, 1987, pp. 252-259. Think also, for example, of New York's municipal infrastructure.

45 M. Granovetter quotes, in his essay "The Idea of 'Advancement' in Theories of Social Evolution and Development" (American Journal of Sociology 85(1979), pp. 489-515), the geochemist Harrison Brown (The Challenge of Man's Future. New York: Viking, 1956) who argues that while, at present levels of technology, energy and other resources will be available almost indefinitely, this is only by virtue of vastly more sophisticated implements than at the beginning of industrialization, when petroleum and ores were available near ground level. If, Brown argues, "this vast technical superstructure were ever destroyed, industrial civilization could never be rebuilt, because without this equipment, energy sources could never again be tapped in sufficient quantity to do so" (p. 507).

CHAPTER 2
THE EMERGENCE OF AN EARLY LARGE-SCALE TECHNICAL SYSTEM: THE AMERICAN RAILROAD NETWORK

Stephen Salsbury

1 Railroads - a break with the past

In his well-known book, *Railroads and American Economic Growth*, Robert Fogel asserted that in 19th century America railroads produced a "social saving" of about "six-tenths of 1 per cent of national income."[1] Fogel's work, in common with that of many economic historians who responded to it, attempts to measure in precise numerical terms the importance of the railroad to America's economy. Nevertheless, regardless of their direct quantitative impact, railroads played a vital part in bringing about and moulding the new industrial world which emerged in the United States after 1840. Railroads were the first large scale technical system which arose in America and as such they shaped the way Americans organized technology and had a profound impact on large scale business. In defining the way in which the United States responded to large-scale technical systems railroads may have their most significant contribution to America's economic growth. This is a contribution that cannot be easily measured.

There is a general agreement that the United States underwent a dramatic transformation in the 19th century. In 1800 its population was a mere 5.3 million. By the end of the century the nature of the population had changed. In 1800 only 322,000 lived in urban territory and most of these were in places of 25,000 or less. In contrast rural population was nearly five million. One hundred years later America's population was more than seventy-six million, of which 30.2 million was urban.[2] Both the population growth and its composition owed much to structural change in the United States' economy. In 1900 most of the nation's people farmed the land. Commercial banks were just beginning to be formed and the country had no investment banks or stock exchanges. Nor were there any modern factories or industrial corporations. These only began with the textile revolution in 1812. By 1900 big scale industry was normal in many fields, including textiles, meat packing, iron and steel, coal mining, oil refining, and the explosives industry. Many of these industries did not exist and were not even imagined in 1800. Railroads played a crucial part in this change.

Railroads, when they emerged on the American scene in the 1830s, represented a sharp break with the past. They were much larger enterprises than any previously known. Although most started small, they quickly dwarfed the canal systems which at the beginning of the 1830s were the United States' biggest organizations. For example the Erie Canal, completed in 1825, was 363 miles in length and cost $7,000,000. By contrast the small Western Railroad of Massachusetts which opened in 1842 had only 160 miles of single track and cost in excess of $7,000,000. In 1854, although the Western operated no more mileage, its capital amounted to $10,000,000. By 1860 the New York Central Railroad which competed with the Erie Canal had a capital in excess of $30,000,000 and by 1883 the Central, with a trackage of 953 miles had absorbed nearly $150,000,000. A decade earlier the rival Pennsylvania Railroad had a capital of nearly $400,000,000.[3] Prior to the Civil War the nation's largest industrial ventures, the textile mills, were minuscule when compared to the railroads. Few such corporations had a capital of $500,000 and most were capitalized at figures below $250,000.[4]

Railroads were a sharp break with the past in other ways and much of their uniqueness resulted from their technological requirements. Railroads, unlike rival transportation methods, were integrated enterprises. Whereas canal and turnpike authorities built infrastructure that others used and steamboats plied waters provided by nature, railroad corporations built their rights of way, maintained them, and operated all the vehicles used upon them. Furthermore the operation of trains required careful managerial structures of a kind not yet seen in the United States.

It must be emphasized, however, that when railroads first made their appearance their promoters had no idea that their creations would be different. The first American railroads were chartered in the late 1820s and early 1830s. While it is true that railroads are an English invention, America was quick to adopt the idea. The first two important English common carriers, the Stockton & Darlington and the Liverpool

& Manchester were opened in 1825 and 1829 respectively. Maryland chartered what turned out to be one of America's leading railroads, the Baltimore & Ohio, in 1827. The B&O opened its first tracks in January 1830 and the first trains were drawn by horses. This is not surprising since the B&O had been chartered and construction had started before 1829, the year England's famous Rainhill trials demonstrated the practicability of Robert Stephenson's steam locomotive, the Rocket.[5] Well after 1829 American states continued to charter railroads with the assumption that they were little different than turnpikes. In 1831 Massachusetts authorized the Boston & Worcester Railroad and its charter was ambiguous about who would control traffic over its tracks. Many of the state legislators who voted for the charter assumed that the line would be operated by horsepower and that it would be possible for private individuals to put vehicles on the line in the same manner that they did on a turnpike.[6] American railroads emerged contemporaneously with their British counterparts and each nation's lines evolved traditions independently of the other.

There was no grand Federal scheme for transportation development in the United States. Thomas Jefferson's Secretary of the Treasury, Albert Gallatin, submitted a comprehensive plan for the building of a national turnpike and canal network in 1808, but this foundered on the rock of financial austerity caused by America's increasing involvement in the Napoleonic Wars.[7] After the end of the War of 1812 Jefferson's successors, Madison and Monroe, had doubts about whether the Constitution allowed the Federal Government to participate in the building and operation of canals and highways. Consequently such projects were left to the states. By the time railroads became an alternative form of transportation many states were running into financial difficulties with over-ambitious canal projects. Thus the vast majority of the early railroads resulted from private efforts which were sometimes undertaken in cooperation with local cities or occasionally in cooperation with state governments.

American railroading in the 1830s and 1840s was characterized by dozens of locally promoted short lines which were constructed separately from each other. Boston provides an extreme example. Radiating from the city from south, to west, to north were the Old Colony, Boston & Providence, Boston & Worcester, Fitchburg, Boston & Lowell, Boston & Maine, and Eastern. None of these lines were more than sixty miles long and all depended on connections to reach important traffic sources.

For example the Boston & Worcester came to rely on the Western Railroad to tap the Erie Canal at Albany. The Fitchburg required two connections to reach the Erie: the Vermont & Massachusetts and the Troy & Greenfield. In New York State it was much the same story. The Erie Canal was paralleled by no less than ten short lines between Albany and Buffalo. Even lines with grandiose aspirations such as the Baltimore & Ohio in the beginning were more like short lines than the trans-sectional routes they hoped to become. The B&O did not reach the Ohio River at Wheeling until January 1853, and even then the line was less than 400 miles in length.

Because America's railroad system in the 1840s was composed of dozens of different small lines with independent managements it would have been difficult to predict that a national network would arise. That one did was due to two factors. First were the technical requirements of railway operations. Second, political and economic factors within the American Republic forced the formation of a nationwide system.

2 The growth of large organizations

The requirement for the railroads to own and control all vehicles upon their lines brought into existence complex bureaucratic managerial structures. Alfred D. Chandler, Jr. has described this process in detail and it is only necessary to summarize his findings briefly here.[8] The Boston & Worcester, a pioneer American Railroad, which went into full operation in 1835, did not divide its management into functional departments, nor did the B & W have complex operating arrangements. This is demonstrated by the way trains were controlled. It took only about two and one half hours for trains to run from one terminal to the other and to avoid accidents the railroad constructed a passing track halfway between Boston and Worcester and started trains out from each terminal at the same time. The first train to arrive at the passing track was required to wait until the other train arrived. This arrangement did not require any signaling system or even an especially strict supervision of the operating men.[9]

The organizational structure of the Boston & Worcester could not serve a longer, more complex system such as the 160 mile Western

Railroad, which was completed in 1841. For a brief time the Western was the United States' busiest rail line.

Nevertheless Western Railroad did not adopt a new managerial structure until a series of disastrous wrecks, the most serious of which occurred during the first months of full operation.[10] The Western Railroad's response to its safety record was to institute a new managerial structure. The goal was to create clear lines of authority and responsibility. Thus they broke the Western into three divisions, each about fifty miles long. This made it possible for a division officer to supervise personally the things for which he was responsible. The president of the railroad was the system's chief administrator. Beneath him were functional officers with appropriate subordinates on each division. A Chief Engineer (later the title became Superintendent) was put in charge of the railroad's structures and track. He in turn directed a roadmaster on each division. Roadmasters had direct responsibility for keeping bridges, track and roadbed in good repair. Each roadmaster kept a written record of his work and made formal monthly reports to the Chief Engineer. A Master of Transportation controlled train movements; he in turn supervised deputy masters of transportation on each division. A Master Mechanic was responsible for repairing rolling stock at the system headquarters in Springfield, and he in turn had a subordinate in each division in charge of keeping these locomotives and cars in good running order. The Western Railroad's new plan of organization was the first functionally departmentalized, decentralized managerial structure on an American railroad and at the time of its adoption was unique in American business.[11]

As Alfred D. Chandler, Jr. has demonstrated the emergence of long trans-sectional railroads in the 1850s spurred the development of the new administrative methods pioneered by the Western Railroad. The leading systems were the Baltimore & Ohio, Erie and the Pennsylvania, all of which by 1853 had crossed the Appalachian Mountains and reached the Great Lakes or the tributaries of the Mississippi River. The Erie's Daniel McCallum and the Pennsylvania's J. Edgar Thompson were leaders in perfecting a line and staff, decentralized managerial structure. This managerial form was characterized by a general headquarters. There a president supervised several vice presidents, each of whom was concerned with a specific function such as finance, transportation, maintenance of equipment or maintenance of way. As on the Western Railroad these new systems were divided into operating divisions, each controlled by a superintendent. The superintendents were responsible to the president. The functional vice presidents at the railroad's headquarters did not have line authority over the operating divisions. Instead they planned and developed the standards and the rules which the operating officers were expected to follow. The staff vice presidents also received reports from the operating divisions from which they could assess the performance at the operating level. From time to time, they, or their staff, would inspect the divisions. The functional vice presidents did not order changes, but presented their reports to the railroad's president who took the appropriate action. By the 1870s the Pennsylvania's line and staff structure, with its decentralized divisions was in place and the railroad's administrative structure changed very little for the next ninety years.[12]

While most large American railroads adopted managerial systems that resembled the Pennsylvania's, not all did. The New York Central developed a highly centralized managerial form. As in the case of the Pennsylvania, the New York Central divided itself into divisions, each about fifty miles long. While the New York Central had a general headquarters, its functional vice presidents were not staff officers. They had direct authority and control over their functions on the divisions. Although this arrangement gave the officers at the general headquarters less time to plan and analyze (since they were directly involved in the day-to-day running of the railroad) it nevertheless worked reasonably well. In fact once the New York Central perfected its managerial system in the 1870s, the Central, like the Pennsylvania, saw its structure last for some ninety years.[13]

3 Obstacles to system growth and innovative solutions

While it is easy to devise new managerial structures it is not always easy to make them work. The new bureaucratic systems, whether of the Pennsylvania's line and staff, decentralized type, or the highly centralized New York Central kind, encouraged the development and use of new technology. The most critical early problem was the operation of trains. The difficulty was how to convey orders over a far-flung system. It should be emphasized that early American railroads operated largely single track lines and had no signaling systems. It was one thing to print timetables, but quite another to see that they were followed. The problems were many. What happened when a train did not arrive at its scheduled meeting place? What should the opposing train do? Was the late train just a few minutes behind schedule, or had it failed to start from its terminal, or had it broken down at some point and was unable to move? Without a clear procedure that all employees knew and followed the railroad risked chaos and danger.

The Western Railroad, as did other American railroads, developed a rule book which their employees were expected to memorize. These volumes, which often exceeded 100 pages, were supposed to have a solution for any problem. Thus if a train broke down the conductor was to send a brakeman for assistance. The brakeman was expected to find a horse to help him on his way. If the train was scheduled to meet another, the conductor was to send a brakeman forward to meet the oncoming locomotive and to warn its engineer and then the brakeman was to return to his own train with the oncoming train. If a train were following the conductor was required to send a brakeman to meet it.[14] These crude rules were designed to prevent head-on and rear end collisions, and to allow trains to move forward with safety past normal meeting places. The rules did not create a system where central dispatchers could efficiently route a train over a complex system. What was needed was a method to transmit quickly information to central dispatchers, who could in turn send out orders to trains in route.

The electric telegraph met this need. Samuel F. B. Morse's experimental American telegraph was constructed along the Baltimore & Ohio Railroad's Baltimore to Washington branch in 1844. Many observers of the telegraph in the 19th century have observed that the invention had a natural affinity with the railroad. Morse certainly recognized it, since building his line alongside the railroad provided him with a cheap right of way. It would have been expensive to purchase an exclusive right of way, even if the telegraph had been given the power of eminent domain. Interestingly, even though the B&O received the right to use the telegraph for nothing, the system's managers regarded the invention with suspicion. The B&O's management did not make any use of the telegraph to control train movement even though telegraphs had previously been constructed in England where railroad managers made immediate use of them for running trains.[15]

The Erie Railroad, one of the pioneers in administrative innovation, also pioneered in the use of the telegraph. Charles Minot, the Erie's superintendent witnessed Ezra Cornell's attempts in 1849 to construct his New York & Erie Telegraph. Minot suggested that the line be built on the Erie's right of way. Despite the initial objection of the Erie's Board of Directors, Minot had his way and by 1851 the railroad had telegraph wires beside its track. Minot himself issued the first telegraphic order to control trains in June 1851. The superintendent was aboard the Day Express when it came to Turners, New York, where it was supposed to meet another train. The opposing train did not arrive. Minot telegraphed the next station and learned that the oncoming train had not yet appeared. Minot then telegraphed Turners and told the station agent to hold the oncoming train. The Superintendent then issued an order that his own train move forward. The engineer would not obey the order, and Minot himself drove the locomotive to the next station.[16]

It fell to Minot's successor Daniel McCallum to exploit fully and advertise the importance of the telegraph to railroading. McCallum wrote in the Erie's annual report of 1855 that "A single track railroad may be rendered more safe and efficient by a proper use of the telegraph than a double track railroad without its aid... It would occupy too much space," he went on, "to allude to all the practical purposes to which the telegraph is applied in working the road; and it may suffice to say that without it, the business could not be conducted with anything like the same degree of economy, safety, regularity, or dispatch."[17]

The railroad not only used the telegraph, but undoubtedly hastened its spread across the United States. In fact a large portion of telegraph companies were brought into existence by railroads. Thompson, in his book, *Wiring a Continent*, describes in detail a contract between the New York & Mississippi Valley Printing Telegraph Company (which in 1856 became the Western Union Telegraph Company) which in 1854 built a line between Detroit and Chicago using the rights of way of three companies: the Cleveland & Toledo, the Michigan Southern, and the Northern Indiana. Under this contract the railroads built the telegraph to the exact specifications of the telegraph company. The railroads not only gave their rights of way, but they provided the capital. For this they received stock in the telegraph company at the rate of $125 of stock for each mile constructed. The railroads also had free use of the telegraph lines for railroad business. The telegraph company agreed to give priority to all messages connected with train movements.[18]

The Western Union, which eventually controlled most of America's telegraph network, owed much of its rapid rise to a symbiotic relationship with the railroads. Robert Thompson concluded that "to no small degree the future greatness of Western Union was built upon dozens of exclusive railroad contracts drawn up by its founding fathers."[19] Certainly the great railroad entrepreneurs of the mid-19th century recognized the importance of the telegraph. Cornelius Vanderbilt acquired control of the New York Central in 1867 and in 1869 he bought control of Western Union. In 1870 the telegraph company had 12,600 offices of which 9,000 were in railroad depots.[20] In the 1870s the Western Union's strong position was far from secure. The Pacific transcontinental railroads, the Union Pacific and the Central Pacific, were required by their charters to build a telegraph line along their tracks. They turned this over to the Atlantic & Pacific Company which later fell into the hands of Jay Gould who came to control the Union Pacific in the 1870s. This set the stage for a battle between the Gould and Vanderbilt interests for control of the nation's telegraph system.[21]

The often colorful fights between titans such as Gould and Vanderbilt obscure the fact that railroads, because of their growing reliance on the telegraph, brought this revolutionary form of communication to every town and hamlet in the United States with a railroad station. Thus Western farmers, shopkeepers, and bankers were provided instant communication with the great urban centers in the East and by using the Transatlantic cable they had rapid communication with Europe and even Australia. Railroad stations housed most of America's telegraph offices, and railroads also supplied most of the telegraph operators. It is doubtful that telegraph companies would have opened small town offices except to fulfill their obligations to the railroads in the dispatch and control of trains. Thus America's first large scale technical system proved essential in spreading the use of new communication technology.

The telegraph was related to another change brought about by the railroad, a revolution in time. Civilization has long had clocks, but not until the 19th century was there a concept of "standard time". In fact "standard time" was a railroad invention. Timetables were essential in running railroads. They became a necessity for passengers, especially since early railroads serving urban centers such as Boston started to run frequent trains for commuters. Timetables were even more important for dispatching trains and keeping an even flow of activity across a system. Traditionally cities and towns measured time by the sun, and noon was when the sun was directly overhead. Consequently most cities and towns ran on different times - a situation which posed a severe problem for a railroad timetable.

In order to avoid confusion, railroads found it essential to develop a single time standard that would be recognized throughout a region. In 1848 several New England lines formed the New England Association of Railroad Superintendents. In 1849 this organization decided to adopt a single standard time which was to be fixed on Boston time and provided by William Bond & Son. William Bond was the first director of Harvard's Cambridge Observatory. Through Bond's influence New England railroads came to adopt the time provided by the observatory. In 1851 Bond began to telegraph time signals from the observatory. These reached many railroad stations simultaneously and provided the basis for an accurate standard time throughout New England.[22]

Standard time was essential for railroads. In this sense steam trains differed from their competing modes (horses and wagons, stage coaches, canal boats and steamboats) where precision timetabling was not vital or even necessary. The United States, which was very large compared to a European nation, could not adopt a single time for the entire country. Railroads, therefore, took the lead in dividing the American continent into time zones and on November 18, 1883, 600 United States railroads dropped fifty-three arbitrary times they had operated under and adopted a standard railroad time with the four time zones that are so familiar today. This was far in advance of government action. Congress did not formally establish standard time until 1918![23]

Railroads also took the lead in revolutionizing the procurement of sophisticated technical machines. Nothing better demonstrates this than locomotive purchases. Here again railroads were a sharp contrast with the past. Railroads quickly became big organizations with bureaucratic structures that collected information which could be, and was, analyzed at their central headquarters. Locomotives quickly posed a major problem for the new American systems. Early New England lines such as the Boston & Worcester, Boston & Providence, and the Boston & Lowell all imported English locomotives built by Robert Stephenson. During the 1830s English engines and American-built locomotives based on English designs set the pattern.[24]

American railroads differed considerably from the English. As a whole, United States lines had access to less capital and were built more cheaply than their English counterparts. Furthermore in many cases, especially mountainous New England, Maryland and Pennsylvania, American geography was much less favorable for railroad building than in Old England. Consequently most United States lines were built with lighter rail, had sharper curves and steeper gradients than British railroads. This caused American railroads to experiment and in the early years the Baltimore & Ohio served as the nation's railroad "university". The B&O in 1830 tested Peter Cooper's famous "Tom Thumb", whose vertical boiler made the engine look and perform differently from Stephenson's engines which had horizontal boilers. During the very early years upright tubular boilers characterized B&O locomotives such as the "York" and the "Atlantic". They were specifically designed to operate on the system's sharp curves and steep hills. The Baltimore & Ohio soon developed its own shops influenced by men like Ross Winans who had helped in the designing of the successful "Atlantic" in 1832, which unlike the Stephenson engine the "Herald" bought by the rival Baltimore & Susquehanna Railroad did not constantly derail on the tight curves.[25]

The Western Railroad was an early system to encounter locomotive troubles. This line crossed the Berkshire Mountains, and had a summit nearly 1,400 feet above sea level. The mountainous half of the line contained many steep inclines and sharp curves. The Western had from its start used English-designed locomotives. However, when the mountain section opened the company's directors decided to place an order with Baltimore's Ross Winans for seven "crab" engines of the design pioneered by the B&O. These had extremely short wheel bases to go around the 400-foot radius curves which characterized some of the mountainous portions of the B&O. The Western's Chief Engineer George Whistler was enthusiastic about the crab engines. Unfortunately the locomotives proved failures, but what is important is the way the Western came to organize its motive power purchases. About the same time as the order for the crabs, the company bought engines constructed by several other manufacturers including Lowell Massachusetts' Locks and Canals Company, Boston's Hinkley and Drury works, and Philadelphia's William Morris. The railroad put all of its various locomotives through rigorous tests and had its engineers and master mechanics fill out detailed reports each time a machine was in service. The details which flowed into the Western's headquarters included the number of miles run, speed attained, tonnage hauled, fuel consumed, oil consumed, number of breakdowns and the reasons for each, the amount of time the engines were under repair, the cost of repairs, etc. These data allowed the company's executives to make informed decisions on purchasing complex, technical machines. Furthermore most railroads began to build some of their own locomotives since the same machinery that repaired steam locomotives could be used to build engines as well. This made economic sense because it allowed a company to achieve full use of its repair shops and labor during slack periods. Building engines allowed companies to experiment with new designs and to develop an engineering staff that could draw up specifications for locomotives purchased from independent manufacturers such as Baldwin, Norris or others. Some railroads such as the Pennsylvania and the Norfolk & Western ended up building large shops and constructing a large portion of their motive power. Others purchased most of their equipment, but all large systems became expert at evaluating and testing their locomotives. It was the railroads' large size and bureaucratic organization which enabled systematic analysis of locomotive design. Such opportunities were not available to the relatively small and constantly changing river steamboat companies.[26]

4 Network integration and its prerequisites

The American railroad network started with many small companies and grew into a vast system which at its peak in 1916 had more than 254,000 route miles of track.[27] The great majority of American railroads have been, and still are, private corporations. While the number of companies has shrunk over the years (there were 1,085 different firms in 1920), there were still 635 separate lines in 1957. While most of these were short lines of little consequence in mileage operated, passengers carried or tonnage hauled, there were nevertheless in 1957 about sixty major lines such as the Pennsylvania, New York Central, Erie, Baltimore & Ohio, Chicago, Burlington & Quincy, Great Northern, Northern Pacific, Santa Fe, Southern, Southern Pacific, Norfolk & Western, New Haven and many others. Most of these systems had thousands of miles of track and operated over vast distances.[28] The question which arises is: How could such a large collection of different

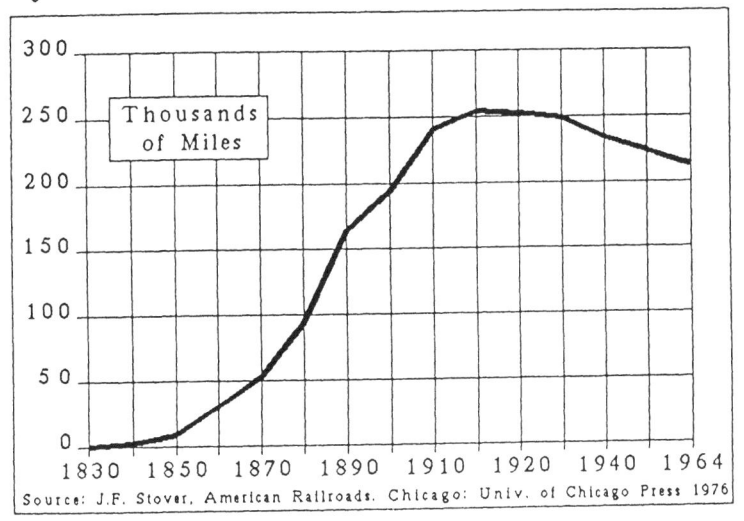

Figure 1: Growth and Decline of Railway Mileage in the U.S.

[Graph: Thousands of Miles, y-axis 0 to 300, x-axis years 1830, 1850, 1870, 1890, 1910, 1920, 1940, 1964]

Source: J.F. Stover, American Railroads. Chicago: Univ. of Chicago Press 1976

companies become by the 1890s a unified network with the same gauge, easy interchange of trains and cars, through freight billing and ticketing procedures, and common accounting practices? So integrated was the railroad network that it was possible for a passenger to ride a single train between Chicago and San Francisco over three different railroads and the passenger might not possibly be aware that he had travelled on more than one line.

The answer lies both in the nature of the railroad as a technical system and political and economic considerations in the United States. In the 1850s four great trunk railroad systems emerged: the Baltimore & Ohio, Erie, New York Central, and Pennsylvania. The first two had had direct government aid - either state or local - at the start. These two systems differed from most of the early American railroads in that they were conceived of from the start as important trans-sectional routes that were to connect an Atlantic seaboard port with the western waters of the Great Lakes or the Mississippi River system. The Erie Canal had already demonstrated the importance of the inter-sectional traffic in agricultural commodities flowing from the newly developing regions of Western New York State and Ohio.[29] The problem for all of the four large trunk routes in the 1850s was that much of the traffic which they sought to carry originated outside the states in which they were chartered. Significantly from the first, railroads in the United States were chartered by the state governments, not the federal government. State-chartered railways in Ohio, Michigan, Indiana, Illinois and other trans-Appalachian states were springing up at the same time the four great Eastern trunk lines were finishing their lines to the Ohio River and Lake Erie. As early as the 1850s it was clear to the leaders of the Baltimore & Ohio, Erie, New York Central and the Pennsylvania that they must make arrangements to coordinate the flow of commerce between themselves and the emerging systems in states such as Ohio and Michigan. As Alfred D. Chandler, Jr. has noted the early response of the eastern trunk lines was to make alliances with their western connections. Thus the Pennsylvania bought into and made agreements with such lines as the Pittsburgh, Fort Wayne & Chicago, and the Indiana Central. The New York Central made agreements with the Lake Shore and Michigan Central systems.[30]

It is often forgotten that the United States started with different gauges. While New England largely adopted the English standard gauge of 4' 8½" (as did the New York Central and the Baltimore & Ohio) the New York & Erie (Erie) was built to a broader 6' gauge. The charter of the Erie contained an interesting clause. Because its New York City promoters wanted to insure that they would be able to monopolize the trade of southern New York State, the charter specified that the system would forfeit its charter should the line connect with any railroad leading into Ohio, Pennsylvania or New Jersey.[31] The fact that Ohio became one of the leading agricultural states, whose products were in high demand on the eastern seaboard and abroad, forced the Erie to either miss out on much traffic or ask for a change in its charter. The lure of opportunity west of New York State was too great to resist and by 1864 the Erie had caused a broad gauge 6' line to be constructed across Ohio to Cincinnati where it met another 6' gauge railroad which ran all the way to St. Louis on the Mississippi.[32] The Erie was not the only railroad to be built to a different gauge. In much of the South (especially South Carolina, Georgia, Tennessee, Alabama and Mississippi) most railroads were of a 5' width. As of 1861 American railroads employed no less than eight different gauges ranging from 4' 3" to 6'.[33]

Initially American railroads in the 1860s were not interested in building a single national system. They merely desired to cater to grow-

Figure 2: Principal Companies of the Seven Major Railroad Combinations in the Early 20th Century

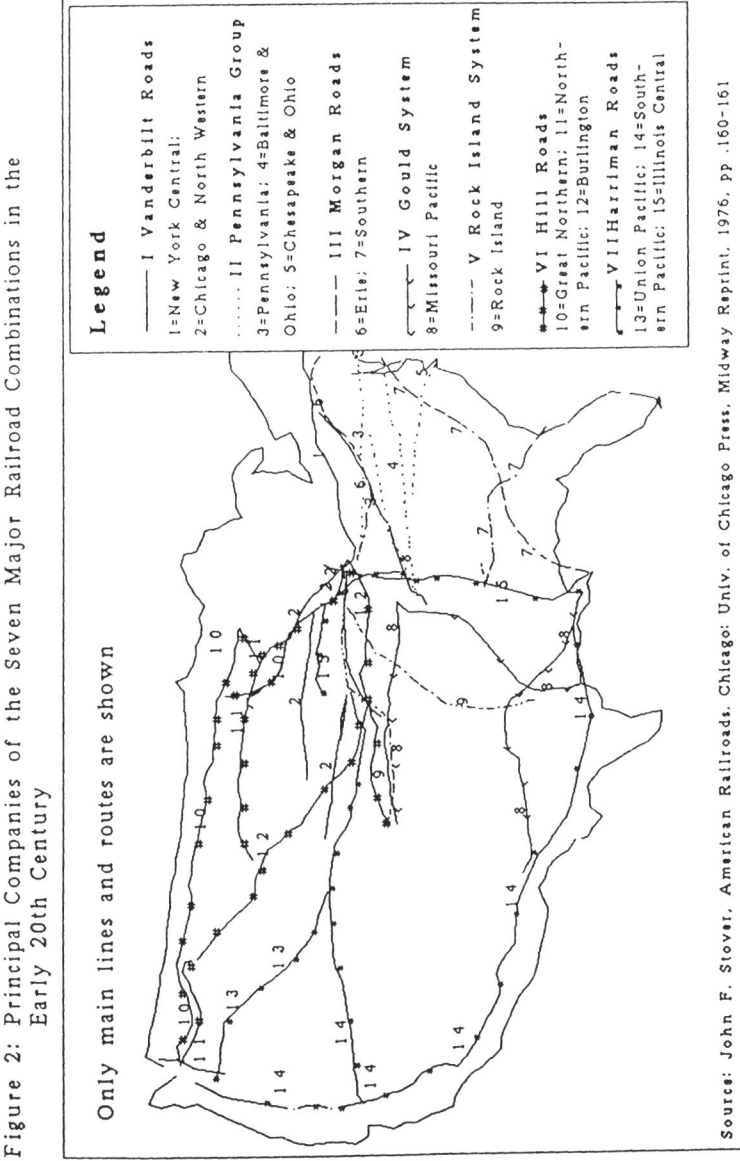

Only main lines and routes are shown

Legend

——— I Vanderbilt Roads
1=New York Central;
2=Chicago & North Western
.... II Pennsylvania Group
3=Pennsylvania; 4=Baltimore & Ohio; 5=Chesapeake & Ohio
——— III Morgan Roads
6=Erie; 7=Southern
⌒⌒⌒ IV Gould System
8=Missouri Pacific
– – – V Rock Island System
9=Rock Island
✳✳✳ VI Hill Roads
10=Great Northern; 11=Northern Pacific; 12=Burlington
——→ VII Harriman Roads
13=Union Pacific; 14=Southern Pacific; 15=Illinois Central

Source: John F. Stover, American Railroads. Chicago: Univ. of Chicago Press, Midway Reprint, 1976, pp.160-161

ing inter-regional traffic between the mid-West and the Atlantic seaboard. Unfortunately the alliance system proved ineffective. Jay Gould, one of the United States most infamous railroad leaders destroyed the usefulness of alliances. Gould came to control the financially weak Erie in the late 1860s. He wanted to put his company into financial health by increasing its share of the trans-Appalachian trade at the expense of the other major trunk lines, especially the New York Central (NYC) and the Pennsylvania (PRR). Thus in 1869 Gould tried to undermine the Pennsylvania's western connections by secretly gaining control over such lines as the Pittsburgh, Fort Wayne & Chicago and the Indiana Central. Gould's actions forced the great trunk railroads to attempt to protect their traffic by adopting a strategy that Alfred Chandler has called "system building". Under system building, the trunk lines sought to own their own tracks all the way to the source of their traffic. In practice this meant that the B&O, Erie, New York Central and Pennsylvania railroads all expanded their lines to connect New York City with Chicago. Furthermore the B&O, NYC, and PRR also linked New York with such cities as Cincinnati, St. Louis, Cleveland, and most other major mid-Western centers.[34] In theory, system building should have diminished interchanges between railroads since the aim was to make certain that once traffic got onto a company's tracks it would move over that system to its destination. In these circumstances one might have expected systems to retain their separate gauges in order to make it hard for goods once loaded to be transferred to another railroad.

Even as the system building strategy was being formulated, economic forces were at work leading railroads to recognize the value of a single gauge. Because the American railroad network was still so fragmented in the 1860s, private firms arose to facilitate the movement of goods over long distances. These were the famous "fast freight lines" such as the National Dispatch Company (which was based on the Vermont Central route from New England to the Great Lakes) and the Red Line between Boston and Chicago. Before the fast freight lines arose shippers often found it difficult to arrange for long distance freight movements. In the early years railroads did not offer joint rates or joint bills of lading and often goods had to be trans-shipped (unloaded and reloaded) at transfer points where breaks of gauge occurred. The fast freight lines solved these problems. In some cases such as the National Dispatch Company private corporations bought special freight cars and arranged to move them over several railroads, thus relieving shippers from this task. Thus shippers had to deal only with a single firm which could fix a definite tariff and arrival time.[35] Sometimes railroads banded together to organize fast freight lines. The Red Line was one of these. In so doing the railroads began to offer superior service to all water routes. Railroads combined speed and ease of handling. For example the water route from Boston to Chicago involved at minimum a steamship or a sloop to Albany, a canal boat to Buffalo, and then a steamboat to Chicago. This meant at least two trans-shipments, probably three. In order to be effective, however, fast freight lines had to move goods in a single car from the point of the freight's origin to its destination. George Rogers Taylor and Irene Neu note that in the 1860s railroads began to capture from the waterways the lucrative grain traffic between the Ohio Valley and the eastern seaboard. They estimate that trans-shipment costs on the Great Lakes-Erie Canal route accounted for 20% of the freight charges. The fact that rails eliminated trans-shipment costs and gave fast, year-round service led to their domination of the grain traffic from the Midwest to the Atlantic coast in the 1870s.[36] The elimination of trans-shipment depended on a single gauge and Taylor and Neu credit fast freight lines as being a major force in causing railroads to begin, starting in the 1860s, to adopt a uniform standard American gauge. The logical standard gauge to adopt was 4' 8½", which in 1861 accounted for about 53% of all the trackage in the United States. This figure includes the important Pennsylvania Railroad whose 4' 9" gauge allowed interchange with the 4' 8½" track.[37] The 4' 8½" gauge was further boosted when the Federal Government specified it as the gauge for the first transcontinental railroad from Council Bluffs, Iowa to Sacramento, California which was completed in 1869.

System building, which mainly occurred after 1870, did not lessen the need to interchange traffic between railroads. None of the original great trunk systems ever extended their systems beyond the Mississippi River at St. Louis. Further north in Illinois, Chicago served as the great terminal city where the great systems of the West met those of the East. In the South, New Orleans and Memphis served this purpose. As the center of America's population and commerce moved westward it became clear that traffic would move from across the Mississippi River and even across the Missouri to the East Coast. One example makes this clear.

By the 1870s the United States had become a nation of ever more specialized regions. New England, in colonial times largely self-sufficient in food, had by the 1850s industrialized. Farms were abandoned and New England bought food from western New York State and later from Ohio, Indiana and Illinois. In 1875 Gustavus Swift, a New England wholesale butcher who purchased live animals railed into Boston from western New York State and Ohio, decided to move westward to Chicago. In Chicago he built packing plants to kill beef raised in the Midwest, although some of the animals came from as far away as Texas. Swift's enterprise depended on vertical integration. Since Swift proposed to serve his wholesale New England markets by supplying beef killed in Chicago he had to guarantee that refrigerator cars were immediately ready to move his freshly killed beef. Initially Swift found that the trunk railroads did not want to haul refrigerator cars. At that time the Erie, B&O, NYC and Pennsylvania were under the influence of Albert Fink, the head of the Eastern Trunk Association, who feared that butchered beef would reduce railroad tonnage as opposed to the traditional method of hauling live cattle in stock cars. At this point Swift benefited from the American competitive system. System building had not removed rivalry for the interregional traffic, nor did the Eastern Trunk Line Association include all railroads able to transport freight between Chicago and Boston. Taking a cue from the concept pioneered by the "fast freight lines" Swift developed his own refrigerator cars, which were owned by his business and prevailed on the Grand Trunk Railroad running through Canada to haul his beef to New England. This tactic forced the other Eastern trunk lines to haul Swift's cars. In the United States competition gave shippers a strong hand.

Swift soon developed a large fleet of refrigerator cars which he continued to own and as his business expanded the cars were sent to nearly every corner of the American railroad network. By the 1890s the big meat packers, Swift and those who followed his example such as Armour, Schwartzschild & Sulsberger, and Cudahy came to own large fleets of refrigerator cars, whose movements were not confined to any one system, but followed commercial needs. Furthermore the cars, in order to earn money for their owners, soon engaged in return hauls. They might carry Chicago beef to New Orleans and then move Louisiana strawberries northward. They also hauled fresh fruits from California to the Midwest and East. As the packing industry grew, it moved closer to its source of supply - to Kansas City, Missouri; Omaha,

Nebraska; and St. Joseph, Missouri. This reinforced the developing trend to ship goods through the great terminals at Chicago, St. Louis, Memphis and New Orleans.[38]

American economic forces put a premium on easy interchange between the various railroad systems, regardless of the goals of the railroad leaders. The movement toward a single gauge gained rapid momentum after the end of the Civil War. At the war's beginning some railroads had tried to solve the gauge problem through specialized equipment such as the "compromise car." This vehicle had wheels with 5-inch surfaces and could run on track as slim as 4' 8½" and as wide as 4' 10". These cars could travel on two-thirds of the country's rail network as of 1861, and most importantly they allowed one of the nation's major trunk lines, the Pennsylvania to interchange with the 4' 10" gauge Pittsburgh, Fort Wayne & Chicago.[39] Another method of encouraging interchange was laying down a third rail, a strategy used by the wide gauge Erie to enable it to exchange cars with 4' 8½" lines from its tracks in Ohio to Chicago. By June 1878 the Erie had its third rail in place over the entire system thus adding another 2% of the nation's track to the standard gauge network.[40] In the 1870s many companies made abrupt changes to the standard gauge. This was comparatively easy because these lines were often broad gauge and could easily install the necessary third rail over a period of time prior to the changeover day. It was relatively simple to convert broad gauge lines to the 4' 8½" width because the original bridges, tunnels, embankments, and clearances would be ample. The last big region to undertake gauge change was the South. In February 1886 representatives of the important southern broad gauge lines, who represented in excess of 13,000 miles of track, decided to convert their 5' lines to 4' 8½" gauge on Monday, May 31, and Tuesday, June 1, 1886. This massive changeover was accomplished at a cost of about $150 per mile which included the expense of converting cars and locomotives.[41] The decision of the American Railway Association's Committee on Standard Wheel and Truck Gauges in October 1896 to fix 4' 8½" as the United States' standard gauge was almost an afterthought. Six years earlier in 1890, with the exception of a few systems specifically committed to narrow gauge lines, nearly all of the United States' railroads were standard gauge.[42]

5 The needs for standardization and uniformity

The impact of the interchanging of cars on American railroads can hardly be overestimated. Interchange forced cooperation among the many different corporations willy-nilly and brought about such diverse innovations as standardized couplings, braking systems and accounting practices.

From the first, American railroads were quick to share technical knowledge, and this was certainly a force in encouraging uniformity. The *American Railroad Journal*, started in 1831 by D. Kimball Minor, was filled with engineering data.[43] The Journal's editorial comments condemned bad operating practices such as those which occurred on the Western Railroad in the 1830s, and thus helped to bring about administrative as well as technological advances.

Whereas the years 1840-1870 saw American railroads concentrate on internal administrative and operating problems the period after 1860 saw an increasing emphasis on solving the issues that arose between the different companies. The role that interchange of cars played in this process is very clear. Often cooperative efforts took the form of associations. The American Railway Master Mechanics' Association was formed in Dayton, Ohio in June 1868 and at least once a year thereafter held conventions. One aim of the organization was to set standards and the proceedings of the meetings were published. Meetings considered such important topics as the adoption of standard car wheel centers, the necessity for accurate gauges, and standard diameters of locomotive driving wheels.[44]

In the 1860s a group of upstate New York railroads formed the Master Car-Builders Association (MCBA) and one of its early tasks was to appoint a committee to prepare a "Dictionary of Terms used in Car-Building." This volume, which finally appeared in 1870, has a revealing preface which begins, "Ever since the general interchange of cars among different railroads, a great deal of inconvenience, confusion and delay has been caused to those who build and repair them by the want of common names for the different parts of cars. One part is known by one name at one place and by quite different names at other places; and, what causes still worse confusion, a term often means one thing on one road and quite a different thing on another. A draw-bar is called a 'pull-iron' in one section, a 'shackle-bar' in another, and in some of the Middle and Southern states it is known by the euphonious name of a 'bull-nose'."[45] The book contains a detailed dictionary of terms together with clear drawings of everything from car wheels to bell-cord fixed-hangers.

In October 1872 the nation's railroads doing a gross business of more than a million dollars a year formed the American Railway Association. This organization undertook many tasks including the development of a Car Service Division, which came to direct and keep track of the interchange of cars on the railroad systems. By the 1920s this division was responsible for more than 2,000,000 freight cars.[46] The American Railway Association also formed a Mechanical Division, which by the 1930s had twenty-four sub-committees to cope with such topics as Brakes and Brake Equipment, Couplers and Draft Gears, Safety Appliances, Specifications for Materials, Tank Cars, and Car Construction to name a few.[47]

Nothing better illustrates the process set in motion for uniformity on American railroads by car interchange than the development of air-brakes for freight trains. The air-brake is normally associated with George Westinghouse, but it is often forgotten that a number of others were working on similar braking systems. Westinghouse first applied his straight-air brake on passenger trains in 1868 by equipping the Panhandle Railroad's Steubenville (Ohio) local with such a system. The concept was particularly attractive since in theory it allowed an engineer to brake an entire train as a unit. Prior to that time brakes had to be set by hand on each car, which was labor intensive, dangerous, and not especially effective in an emergency. From the Panhandle, air-brakes spread to the Pennsylvania Railroad in 1869 and then quickly to other lines. As early as 1876, 37% of all United States passenger cars were equipped with the straight-air brake.[48]

Air-brakes were expensive. It paid to use them on passenger cars, especially since there were comparatively few of them. In 1876 there were less than 16,000 passenger cars as in contrast to hundreds of thousands of freight cars. Furthermore interchange meant different things for passenger and freight service. Railroads carefully planned the consists of their passenger trains. They knew which cars would be used in joint service and which would not. Railroads often banded together to buy equipment for jointly run limiteds. Thus railroad management could keep tight control over passenger equipment.

Freight was different. Cars once loaded might start out on the Pennsylvania and end up a continent away on the Southern Pacific. The West Coast railroad would reload the car and there was no guarantee that it would receive a cargo that would take it back to its home line. Therefore a railroad company might go to a large expense to put air-brakes on its cars only to end up seeing them hauling goods on other systems. Furthermore in order for air-brakes to work effectively a whole train had to be so equipped. In short it did little good for a single or a few railroads to go to the expense of buying air-brakes for their cars, the entire United States railroad network had to take action.

The story of how America equipped its freight cars with air-brakes reveals much about how the pioneer large technical system in the United States solved problems. It turned out to be a complex bureaucratic process. First it was necessary to gain recognition and approval for the product. George Westinghouse recognized this when he placed a train equipped with straight-air brakes at the disposal of the newly formed Association of Master Mechanics which was meeting in Pittsburgh in 1869.[49] However it was the Master Car-Builders Association that was crucial in the adoption of air-brakes for freight trains. For one thing, braking on freight trains, which were long and heavier than passenger trains, was a major challenge. In 1877 the MCBA set up a Committee on Continuous Brakes to study the problem. In 1880 the railroads made the MCBA more effective by changing its rules to provide that member railroads would have voting power according to the number of cars they owned. This linked the association more closely to the will of the large systems and gave more weight to its decisions. In 1884 Godfrey Rhodes of the Chicago, Burlington & Quincy Railroad (Burlington), an enthusiast for air-brakes, became the chairman of the Master Car-Builders' Committee on Continuous Brakes. The MCBA authorized the trial of various braking systems for the purpose of developing standards which could be adopted by the Association for the railroad network. In 1886 and 1887 the Burlington held a series of trials of several brake systems which included the Westinghouse automatic air brake, the American Brake Company's direct buffer brake, Eames's automatic vacuum brake, and several others.[50] At first all these different brakes were found wanting, but Westinghouse was soon able to improve his so that it worked effectively. The MCBA then drew up a standard for automatic brakes that in effect specified the Westinghouse automatic air brake.

The MCBA's action was just the first step. Railroads had to be convinced to adopt the system. As of 1893, 90% of all America's freight cars were without the approved air brakes. At this point the whole issue moved to the political arena, largely at the urging of the growing railway union movement which was concerned about the safety of its members. The issue proved a popular one, and in 1893 Congress passed the Safety Appliance Act which required the railroads to adopt for their freight cars the braking standards specified by the Master Car-Builders Association. This very complicated process, which relied on both private and public bureaucratic structures, was effective in equipping the American freight car stock with air brakes, but over time the process also exacted a high price, which was the injection of politics and governments into the railroad system's technological evolution.[51]

The complex process which came to be adopted by the American railroad network to accomodate technological change applied to other areas as well. Accounting is an excellent example. Alfred D. Chandler, Jr. has noted that railroads were the first American organizations to adopt a complex system of cost accounting. His analysis focuses largely on cost accounting for internal managerial purposes[52], but of equal significance was the issue of setting rates for joint traffic. It was this issue which eventually forced upon railroads a uniform system of accounting.

Few issues in the 19th century aroused more public controversy than railroad rates. Low and equitable rail tariffs were considered vital to the success of individual business corporations and to whole cities and regions as well. Boston, for example, wanted the same rate for bulk produce sent from the Midwest to the Atlantic coast for export as applied for traffic destined for New York City and Philadelphia. Railroads, of course, tried to argue that the cost of doing business should have an important impact in rate fixing. In order to make this argument seriously railroads had to be able to allocate costs properly. The issue of cost accounting arose when railroads tried to set joint tariffs for through traffic. The case of the Boston to Albany route in 1841 is a classic early example. The route was operated by two lines, the Boston & Worcester and the Western Railroad. The Western's part of the 200 mile route was about 150 miles or 75%, while the Worcester's part was 50 miles or 25%. Some suggested that the division of freight rates between the two corporations be set pro rata according to the number of miles the goods were carried. This formula would have divided the charges with 75% going to the Western and 25% to the Worcester. The Western's leaders rejected this idea - they wanted 80% or 90% of the revenue - because they claimed the Western had higher costs. Its trains climbed over mountains, which caused the company to sustain both high capital costs and operating costs. The Worcester did not agree. Its officials claimed that the costs of buying land for rights of way and terminals in Boston far outstripped the cost of building over the Berkshire Mountains. Furthermore they claimed that the B&W had high costs in operating the terminals which were needed to export flour and grain through the Port of Boston.[53]

For decades the Boston & Worcester and Western roads fought over the division of rates on the Boston to Albany route. Nothing was settled. As time went on, it became clear that railroads would have to adopt common accounting procedures to allocate costs between various classes of traffic: passenger, mail and express, and freight, and also between the various types of freight such as bulky, seasonal grain shipments and bulky, year-long coal movements. For years the cutthroat competition between the great railroad systems and the ability of large shippers to play one railroad off against the other and demand rebates, made a shambles of rail rates which were set more by what the traffic would bear than by the costs incurred. This unsatisfactory state for both the railroads and many shippers was one of the main reasons for the rise of government regulation and the creation by Congress in 1887 of the Interstate Commerce Commission (ICC). Thus from the first, one of the major tasks facing the ICC was the development of a common railroad accounting system. Not surprisingly the ICC hired people who had railroad experience, and because of this the cost accounting system that the ICC introduced reflected railroad practices in the 1880s and 1890s. It should be noted that the ICC accounting system was not meant to be a management tool. Its purpose was to allow a rational fixing of rates. However since railroad managements were forced to report their operations according to the ICC formula the ICC accounting method began to influence managerial thinking and as recently as the 1960s Alfred Perlman, who was the President of the New York Central Railroad, favored the ICC accounting system over more modern accounting procedures adopted in the 1950s and 1960s by the Pennsylvania Railroad.[54]

6 Decline sets in

The American railroad network as a vital and progressive large scale technical system reached its apogee in the period between 1900 and 1914. From then on the system gradually declined, however the downward trend was so gradual for so long that many contemporaries hardly noticed it. And there was a paradox. While the system as a whole contracted, technological innovation continued at a rapid pace. At the turn of the century electrification began on a large scale and continued into the 1930s. In the 1930s, 1940s, and 1950s steam motive power was replaced by the Diesel locomotive. Signals became more efficient. The adoption of Centralized Train Control made it possible for one track to do the work of two. Light weight metals reduced the dead weight of passenger trains, and bigger locomotives and heavier rails made it possible to increase the size of freight cars and at the same time lengthen trains. Bigger locomotives and heavier rails also allowed freight train speeds to increase dramatically. Beginning in the late 1960s computer systems started to keep track of freight cars.

Nevertheless, despite technological innovation the railroad system as a whole declined. This was reflected in the decreasing mileage of the national network which from the peak of 254,000 miles in 1916 fell to less than 213,000 miles in 1964. Railroads also lost traffic to other modes of transport. In 1916 they carried 98% of the intercity commercial passenger traffic and 77.2% of the intercity freight traffic. In 1965 these figures were 17% of the passenger traffic and 43.5% of the freight traffic.[55] Debate has been long and sharp over the reasons for the decline. Albro Martin in his provocative book, *Enterprise Denied*, has blamed much of the railroad network's troubles on federal regulation, especially the Interstate Commerce Commission.[56] After the ICC initially failed to regulate railroads, Congress gave the commission more power after 1900. The Elkins Act of 1903 prohibited rebating and in 1906 the Hepburn Act "gave the Commission power to fix maximum rates," and "shifted the burden of proof in rate proceedings from the Commission to the railroads and made ICC decisions effective as soon as they were reached."[57] Martin argues that the ICC regulated railroads at the time when no other enterprises were regulated. Worse yet the ICC began its activities at the beginning of a long term period of inflation. The effect of this, asserts Martin, was to drive capital away from railroads to unregulated business such as the manufacturing of automobiles and other consumer durables.

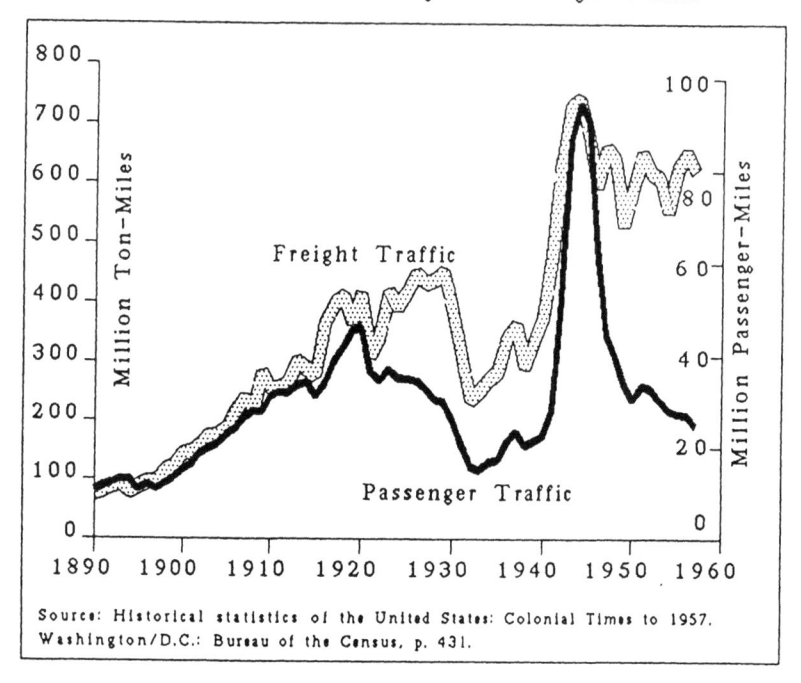

Figure 3: U.S. Railroads: Passenger and Freight Traffic

Source: Historical statistics of the United States: Colonial Times to 1957. Washington/D.C.: Bureau of the Census, p. 431.

Railroad labor unions have received their share of blame for the decline. The unions, which had little success in the 19th century, became ever more important in the 20th century especially under the Wilson Administration during the First World War. Strong labor fixed work rules, which once adopted seemed difficult, if not impossible, to change. Furthermore railroad labor often resorted to government as they had in the 1890s when they helped force the railroads to equip their freight trains with air brakes. Thus during the 1920s and 1930s state after state passed "full crew" laws which fixed manning on trains thus making it more difficult for railroads to take advantage of technological change. Worse yet standard work days based on the number of miles run were fixed in the period before 1920 when train speeds, both passenger and freight, were slow. Thus for railroads both the rise of government regulation and strong labor unions came at a bad time, a time when railroads needed the infusion of much capital and maximum flexibility in order to meet the challenge of rapidly changing technology in water, road, and air transport.

While few would question that the rise of labor and government regulation has hindered railroads, much evidence suggests that another even more basic force was at work causing the rail system to decline. This was bad or ineffective management. At this point it must be emphasized that railroad managers were not bad in the sense of being stupid, corrupt, lazy, poorly educated, or unconcerned with the fate of their enterprises. Rather managers increasingly became the prisoners of the large bureaucratic systems which the railroads themselves had invented in the 19th century, and which for years had worked so well. Large organizational structures create ways of doing things and expectations on the part of those who are employed by them. Some economists have suggested that corporations maximize profits - that is corporations are guided by leaders who invest resources to make the most money. Railroad history does not support this concept. On the railroads, the bureaucrats presided over interests which they sought to preserve. Thus railroad leaders tried to preserve harmony and peace and overlooked ways to make money. Some examples will make this very clear. The giant Pennsylvania Railroad started to electrify some of its main lines prior to the First World War. The new electric locomotives needed no firemen, a fact that was well recognized at the time, even by labor. Nevertheless the Pennsylvania made no attempt to eliminate firemen on electrics. In 1911 a fireman wrote, "It will be a great change for [the firemen] to sit in a nice clean cab equal to a pullman coach, with little more to do than to keep his eyes open, ring the bell for the crossings, and look wise."[58] One does not know why the Pennsylvania did not eliminate firemen from its pioneer electric engines, but I suspect that the system, which had comparatively good labor relations at the time and which was making satisfactory profits, decided not to let technological change upset a carefully worked out harmony.

By the 1920s it was becoming clear to many railroad leaders that their companies were in trouble. The best indication of this was the operating ratio, which records how much of a railroad's total income the expenses of operation consume. In the 19th century most railroads recorded operating ratios which varied between 50% and 70%. That meant that after operating expenses were paid, between half and 30% of a railroad's income could be reinvested, pay interest on the debt, and pay dividends to shareholders. On the Southern Pacific, one of the most powerful of the Western lines, the operating ratio stood at 62.17% in 1917, 85.27% in 1920, and 78.88% in 1921.[59] The Southern Pacific's operating ratio declined a bit during the rest of the 1920s, but shot back up to 80% in 1932. During the 1930s the Southern Pacific began to worry about its ability to service its debt. Southern Pacific's experience was typical of most railroads during the years after 1919.

The ominous upturn of the operating ratio after 1917 did not trigger major changes in railroad management. Nothing better illustrates this than passenger services. In 1934 the Chicago, Burlington & Quincy Railroad brought out its revolutionary Pioneer Zephyr, a Diesel-power seventy-passenger train which also carried a baggage and mail car. This train demonstrated its potential by covering the 1,015 miles between Chicago and Denver in 785 minutes.[60] In theory this train could have operated with a crew of two, or if food was served on board a crew of three. The train could have been made longer and still have had a crew of three. The historian of the Burlington, Richard Overton, commented about the Pioneer Zephyr: "A new age had dawned for the railroads."[61] It had not. Even the Burlington refused to exploit fully its technology in ways that would make the passenger train competitive with the bus and the newly emerging airplane. The new Zephyr fleet had crews that in numbers matched traditional steam trains. Many other railroads refused to even adopt the new technology. Explicitly rejecting the Burlington's lead, the Southern Pacific in 1937 lauched its famous Daylight. This conventional steam-powered train covered the 470 miles between Los Angeles and San Francisco in nine hours and forty-five minutes with a crew of forty-five! This was a ratio of nearly one crew member for every twelve passengers![62]

The fact is that railroad bureaucracy which was so innovative in the 19th century had become rigid and unimaginative in the 20th century. Nothing better illustrates this than the case of the Pennsylvania Railroad. In the 1950s David Bevan joined the Pennsylvania's top management. His background was in banking and insurance and he was stunned at what he found. During the early part of the 20th century big American corporations began to use accounting as a managerial tool. They used accounting data to project future trends and thus to draw up plans for capital expenditures. Managers were particularly concerned that they would be able to determine which parts of their business were most profitable. Railroads, and particularly the Pennsylvania, did not share in these advances. In contrast their practices became frozen about the beginning of the 20th century. Bevan found the vice-president in charge of finance did not have control over real estate or taxes, nor did he supervise the comptroller of accounting. Furthermore the railroad had no capital and income budgets or cash-flow estimates. The rudimentary budgets that did exist were "made up by the staff of the operating vice-president for his use, and were changed from time to time as he saw fit. They were not generally available to top management." Consequently there was no forward planning as to how to meet maturing obligations or to forecast the need for capital improvements. The accounting department could not determine how many people were really involved in accounting, and not a single person in the 2,700-man department was a certified public accountant! Furthermore a large number (nobody knew how many) were doing accounting work in the operating departments. Significantly the accounting department merely followed the Interstate Commerce Commission formulas. The problem was that the ICC system was passive, that is it recorded facts. It had no built-in requirement for forward planning. Worse yet the railroad had no method of knowing which type of traffic made money and which did not. The top management normally made investment decisions on the basis of traffic volume rather than revenue, and most of its estimates were of the seat-of-the-pants variety, not based on hard statistical data.[63] The Pennsylvania was not alone. The Southern Pacific's historian Don Hofsommer found that in the 1950s the SP's "operating department traditionally calculated what trains it could conveniently and economically schedule, and then the traffic or sales department attempted to sell the service. In no case was the sales force responsible for profitability. An officer of an eastern carrier that was particularly progressive in this regard and who deeply admired SP's remarkable car fleet and its well-known operating skills complained, nevertheless, that SP's management had "little or no comprehension of the economics involved."[64]

The American railroad network, the country's first large scale technical system started its life unaware that it would institute sharp breaks with past business traditions. Technological problems caused early railroad leaders to pioneer in devising new structures to manage their companies. The managers were also quick to adopt new technologies such as the telegraph, and in so doing the railroad system had repercussions far beyond that of the railroad industry itself.

Nevertheless, as the network grew not only did it create large bureaucratic structures to manage it, which had the tendency to ossify, but it also brought into being associations, government agencies and labor unions which began to share in the decision making process. This restricted management's ability to react to new challenges, and drove investment money away from the railroads to less restricted fields. The railroads' early success, and their long positions of power in the American economy only seemed to reinforce managerial structures and make them less able to change with the times. Only with the collapse of the giant Penn Central Railroad system in the 1970s did railroad managers, as well as government and union leaders, begin to question the old managerial system. Symbolic of this were new laws which attempted to deregulate railroad ratemaking. At this point, however, it still remains unclear whether any long term or lasting change has occurred.

Notes

1. Robert Fogel, Railroads and American Economic Growth: Essays in Econometric History. Baltimore/Md.: Johns Hopkins Press, 1964, p. 47.
2. Figures from Historical Statistics of the Unites States: Colonial Times to 1957. Washington/D.C.: Bureau of the Census, second printing, 1961, pp. 7, 14.
3. Alfred D. Chandler, Jr. and Stephen Salsbury, "The Railroads: Innovators in Modern American Business Administration" in Bruce Mazlish (ed.), The Railroad and the Space Program: An Exploration in Historical Analogy. Cambridge: The M.I.T. Press, 1965, p. 129.
4. Ibid., p. 130.
5. John F. Stover, History of the Baltimore and Ohio Railroad. West Lafayette/Ind.: Purdue University Press, 1987, pp. 29-32.
6. Stephen Salsbury, The State, the Investor, and the Railroad: The Boston & Albany 1825-1869. Cambridge: Harvard University Press, 1967, pp. 82-83.
7. This plan is described in detail in Carter Goodrich, Government Promotion of American Canals and Railroads 1800-1890. New York: Columbia University Press, 1960, pp. 19-48.

8 Alfred D. Chandler, Jr., The Visible Hand: The Managerial Revolution in American Business. Cambridge: Harvard University Press, 1977, Chapters 3-6.
9 Salsbury 1967, op. cit., pp. 114-116.
10 American Railroad Journal and Mechanics Magazine, September 15, 1841, p. 161.
11 Salsbury 1967, op. cit., pp. 182-184; Chandler 1977, op. cit., pp. 96-98.
12 Chandler 1977, op. cit., pp. 97-106.
13 Ibid., p. 107.
14 Salsbury 1967, op. cit., p. 188.
15 Stover 1987, op. cit., pp. 59-60.
16 Robert Luther Thompson, Wiring a Continent: The History of the Telegraph Industry in the United States 1832-1866. Princeton: Princeton University Press, 1947, pp. 206-208.
17 Quoted in Ibid., p. 211.
18 Ibid., p. 275-276.
19 Ibid., p. 290.
20 Maury Klein, The Life and Legend of Jay Gould. Baltimore/Md.: John Hopkins University Press, 1986, pp. 196-197.
21 Ibid., pp. 197-207.
22 I am indebted for the information on Bond to a forthcoming article in Technology and Culture entitled: "'The Most Reliable Time': William Bond, the New England Railroads and Time Awareness in 19th-Century America."
23 Ian R. Bartky, "The Invention of Railroad Time", Railroad History (1983) Bulletin 148, p. 13.
24 Edward Chase Kirkland, Men, Cities and Transportation: A Study in New England History. Cambridge: Harvard University Press, 1948, Vol. 1, pp. 303-304.
25 Edward Hungerford, The Story of the Baltimore & Ohio Railroad. New York: B.P. Putnam's Sons, 1928, Vol. 1, pp. 103-113.
26 Salsbury 1967, op. cit., pp. 175-178; 190-193; 233-235; 272-273.
27 John F. Stover, American Railroads. Chicago: University of Chicago Press, 1961, pp. 223-224.
28 Ibid., pp. 220-221.
29 For a good discussion of the response to the Erie Canal see: Julius Rubin, Canal or Railroad? Imitation and Innovation in the Response to the Erie Canal in Philadelphia, Baltimore and Boston. Transactions of the American Philosophical Society, New Series Vol. 51, Part 7, Philadelphia: The American Philosophical Society, Nov. 1961.
30 Chandler 1977, op. cit., pp. 148-150; 156-157.
31 George Rogers Taylor and Irene D. Neu, The American Railroad Network 1861-1890. Cambridge: Harvard University Press, 1956, p. 25.
32 Ibid., pp. 39.
33 Data from railroad maps in the back of Ibid.
34 Chandler 1977, op. cit., Chapter 4.
35 Kirkland 1948, op. cit., pp. 441-442; 500-501.
36 Taylor and Neu 1956, op. cit., p. 67.
37 Ibid., pp. 14; 81.
38 Chandler 1977, op. cit., pp. 301-311; 391-402.
39 Taylor and Neu 1956, op. cit., pp. 59, 14.
40 Ibid., pp. 62.
41 Ibid., pp. 79-80.
42 Ibid., pp. 82.
43 Stover 1961, op. cit., p. 21.

44 George L. Fowler (compiler), Index of the American Railway Master Mechanics' Association Proceedings, from Vol. I to Vol. XXXIII, Chicago: Henry O. Shepard, 1901, pp. 199-206.
45 Matthias N. Forney, The Car-Builder's Dictionary: An Illustrated Vocabulary of Terms which Designate American Railroad Cars, their Parts and Attachments. New York: Railroad Gazette, third thousand, 1881, p. iii.
46 Anon. The American Railroad in Laboratory. Washington/D.C.: American Railroad Association, 1933, pp. 9-10.
47 Ibid., pp. 43.
48 Henry G. Prout, A Life of George Westinghouse. New York: American Society of Mechanical Engineers, 1921, pp. 29; 32-33.
49 Ibid., pp. 30.
50 Ibid., pp. 52.
51 This section on the air brake owes much to the excellent essay by Steven W. Usselman, "Air Brakes for Freight Trains: Technological Innovation in the American Railroad Industry 1869-1900," Business History Review, (1984) Spring, pp. 30-50.
52 Chandler 1977, op. cit., pp. 115-121.
53 Salsbury 1967, op. cit., pp. 223-244.
54 Stephen Salsbury, No Way to Run A Railroad: The Untold Story of the Penn Central Crisis. New York: McGraw Hill, 1982, pp. 118-119.
55 Stover 1961, op. cit., pp. 224, 238.
56 Albro Martin, Enterprise Denied: Origins of the Decline of American Railroads, 1887-1917. New York: Columbia University Press, 1971.
57 Thomas K. McGraw, "Regulatory Agencies" in Glenn Porter (ed.), Encyclopedia of American Economic History, Vol. 2. New York: Charles Scribner's Sons, 1980, p. 797.
58 Quoted from the Brotherhood of Locomotive Firemen of February 1911, in Michael Bezilla, Electric Traction on the Pennsylvania Railroad 1895-1968. University Park/Penn.: Pennsylvania State University Press, 1980, p. 135.
59 Don L. Hofsommer, The Southern Pacific 1901-1985, College Station/Tex.: Texas A&M University Press, 1986, p. 113.
60 Richard Overton, Burlington Route: A History of the Burlington Lines. New York: Alfred A. Knopf, 1965, pp. 395-396.
61 Ibid., pp. 397.
62 Hofsommer 1986, op. cit., p. 136.
63 This is condensed from Salsbury 1982, op. cit., pp. 50-51; 189.
64 Hofsommer 1986, op. cit., pp. 283-284.

CHAPTER 3
THE EVOLUTION OF THE TECHNICAL SYSTEM OF RAILWAYS
IN FRANCE FROM 1832 TO 1937[1]

François Caron

The chief characteristic of the history of railways in France is the speed with which a coherently organized operational system was set up. This system was based on the following guiding principles:

1. Lines were granted to private companies for a limited period. Up to 1852, this was very variable but thereafter was fixed at 99 years, as a general rule.

2. The lines were grouped into homogeneous geographical zones so as to form a series of réseaux (networks), each run by a company which enjoyed a monopoly. Railway geography was not mapped out permanently until the signing of the 1883 conventions but its overall pattern was traced by 1859. The networks established under this system were the Ouest, Nord, Est, Paris - Lyon - Méditerranée, Paris - Orléans, Midi and, lastly, Etat. The Etat network lay between the Paris - Orléans and the Ouest and was only formed in 1883. It was not granted as a concession but was managed by a state enterprise. In 1908, the Ouest network was taken over and became part of the Etat. Marking the frontiers, in each case, was the outcome of a complex process that involved not only the interests of the concessionary companies but also those of the districts and towns the networks served.

3. The relationships between the conceding State and the concessionary companies were set forth in a series of agreements confirmed by laws. These agreements were not properly harmonized, however, until 1883 and even then only partially. In the meantime, the terms of the concessions differed widely. The principal feature of the 1883 conventions was that they provided for the application of the interest guarantee. This scheme had been introduced as early as 1839 for the Paris - Orléans and was used after 1857 for lines recently constructed or granted and made up the "new network". In 1883 its application

became general. Under its conditions the companies had to accept a twofold form of state tutelage: on the one hand they were bound by a *"Cahier des charges"*, which listed their responsibilities and liabilities towards government. This measure was finalized in 1847. They were also subjected to a financial control that entailed close administrative supervision. Their freedom of action was curtailed, as both the rates they applied and the investment expenses they incurred had to go through a procedure of authorizations.

4. Thanks to the interest guarantee, the funding of the construction and investment costs which were designed to cater to the growth of traffic was ensured to a very great extent (about three-quarters in 1883) by the issue of bonds. These were in popular demand precisely because of the guarantee and the French railway network could consequently be extended indefinitely.

These four aspects of the French railway system did not become permanent features until after the 1883 conventions, with the removal of the friction which had been mounting between the companies and the State from the very beginning. The distinctive charachteristics of the system, however, were quite recognizable by the late 1850s. It was then, to quote Louis Girard, that the railways ceased to be primarily a "speculation" and became "an institution".

The approach adopted here to describe these developments is historical. First a formative period will be described, characterized by the cautious steps of the actors in the face of a hazardous technical future. The days of trial and error ended in the 1850s when the institutional system achieved an equilibrium, albeit fragile, that could no longer be challenged whereas the technical system proper attained a degree of maturity that ensured its efficient working in economic terms.

Yet, as will be shown in part two, it was not long until railway technology, having crossed its first hurdles, entered upon a period of crisis in the late 1860s under the pressure of a rise in traffic due to the effects of outside factors linked to the operation of the system. In an attempt to answer the new needs, railway technology underwent a transformation of such magnitude that the years between 1870 and 1885 might almost be called a renaissance. A second technical system made its appearance while the old organization of the 1840s and 1850s, if anything, was strengthened.

The evolution of the French system reached its peak in the 1890s at the very time when Kaufmann[2] drew up his laudatory report. But before the First World War that balance was threatened. The ensuing crisis was to lead to the nationalization of 1937, when it became quite clear that the directors of the networks were no longer able to control their costs.

The directors deemed it best to let the State manage the deficits as they held it chiefly responsible for their difficulties. The advocates of nationalization, for their part, placed their faith in the virtues of a technical unification of the networks as a workable means of pulling the railways out of the crisis.

1 The first railway system: 1832 to 1859

Point of departure: 1832

The point of departure in the history of the French railways was the year 1832. Four events in that year marked a decisive and irreversible turning point in technology, ideology, institutions and law. In 1832 Marc Seguin conducted his first experiments in the use of the locomotive with a tubular boiler, trying out his improvements in 1829 on the line from Saint-Etienne to Lyon, which had been a concession since 1826. At the same time and on the same route the first experiments in passenger transportation were carried out. Both acts were symbolical because they opened new prospects for a technology that hitherto had been no more than a tributary of the canal and had only used horsepower.

In 1832 again, several articles and pamphlets appeared, putting forward ambitious projects for the construction of a coherently organized system and making a deep impression on enlightened minds: first, the economist Michel Chevalier (1806 - 1879), a disciple of Saint-Simon and future councillor to Napoleon III, in the newspaper *Le Globe*. Next, four engineering scholars, two from the Polytechnic, Georges Lamé (1795 - 1870) and Emile Clapeyron (1739 - 1864) and the Flachat brothers, Stéphane and Eugène (1802 - 1873) published their "political and practical view of public works in France". Finally, Emile Pereire (1800 - 1875), the great financier, like his brother Isaac (1806 - 1880), voiced the need, in the newspaper *Le National*, to enlist the support of the banking world alongside the efforts of the State in order to launch the railways and applied his theories spectacularly and symbolically by requesting the concession of the line from Paris to Saint-Germain in September 1832.

Meanwhile, the State was far from idle. Still in 1832, Adolphe Thiers succeeded in having the first state subsidy for the railways voted in parliament. It amounted to a grant of 500,000 francs and was used by the Department of Ponts et Chaussées (Highway Department) to evolve a plan for a network throughout the country. Masterminded by Alexis Legrand, the Director General of the Department, the plan was laid before parliament in 1837 and incorporated in the law of 1842. Through his gesture, Thiers had eloquently demonstrated the will of the State to control the process by subordinating individual projects to a guiding rationale. In 1833, the State displayed its will to take charge of the railway construction program even more strikingly by transferring the power to declare public utilities from the government to the legislative. Concessions in perpetuity were excluded altogether and replaced by temporary concessions.

The possibility of building railways on a large scale consequently appeared to be materializing and the desire to do so was loudly proclaimed. Yet many years were to pass before that possibility could take shape and those desires be translated into significant acts.

In order to understand the delays and tergiversations it is important to be acquainted with the principal actors in the game being played round the concession system and the construction of railways and to try to grasp the extent of the uncertainties such an undertaking continued to harbour for a very long time. Three groups of actors were involved: engineers eager to create, senior executives concerned with safeguarding the rights of the State and lastly, the Parisian bankers, anxious to venture only where wise, yet at the same time unwilling to let an opportunity slip to make what might be a sizeable profit. The hesitation on all sides was due as much to the doubts which accompanied the introduction of the new technical system as to differences of motivations or conflicts of interests.

Three actors - three motivations

The opinion campaign waged in 1832 in favor of setting up a national railway network associated pure engineers like Lamé, Clapeyron and Flachat with businessmen like Paulin Talabot and Emile Pereire. Some were from the Polytechnic (Lamé, Clapeyron, Talabot), all were guided by a progressionist and nationalist ideology and sought above all to create an operational instrument capable of increasing their social influence out of all proportion. Seen in that light the railways took on symbolical meaning: they laid the foundations of a new power based on the command of knowledge. Etienne Flachat, who played a considerable role in the conception and completion of the new technical system was also the founder of the Society of Civil Engineers in 1848. But none of them alone could ever have set up the new technical system. Even their coalition was not enough. To succeed they had to convince senior executives and great bankers whose motivations sprang from different if not conflicting sources.

The Department of Ponts et Chaussées disposed of a coherent tradition and doctrine concerning routes of communication which allowed it to tackle the question of the railways firmly and energetically. Its aim was to pursue the task commenced in the eighteenth century, and to construct a well ordered system of transportation governed by the central authorities, in other words by itself. For roads and waterways, networks were extended and infrastructures improved and also a remarkably efficient system of high speed transportation over long distances was introduced. The planning of the connections and even of the traffic, as well as the rates, on the first lines owed a great deal to this experience.

The Department also had pragmatic reasons for wishing to build an extensive railway network. On some sections of the roads and navigable waterways the pressure of traffic had increased steadily throughout the 1830s and 1840s, giving rise to overcrowding that was harder and harder to control. The Department was therefore justified in regarding the construction of a railway network as a rational answer to such problems. Indeed it seemed like the natural outcome of the incapacity of the previous system to fill stated or latent needs.

By persevering in its intention to build (or have built) a coherent and centralized network the Department was also expressing the aspirations of contemporary public opinion, which was impregnated with the general desire to travel. The note supporting the request for the aforementioned grant of 500,000 francs is illuminating in this respect: the primary purpose of a railway was "to provide passenger transportation because, if the railways are well made they will be the sole means of ensuring movement at the highest speed with the greatest economy..."[3].

The desire to travel was an essential component of the Romantic Age and it is important to bear in mind the variety of forms it might take. There were the businessmen with vested interests at home or farther afield and tourism attracted by a growing number of new places and new cities. There were harassed suburbanites, commuting daily, city dwellers in search of Sunday escape and temporary or permanent migrants reluctant to lose all ties with their families and place of origin.

Where the railways were concerned, however, the Department was obliged to adjust to specific technical realities, to pay heed to the state of public and parliamentary opinion and to take financial facts into account. Maintenance activities on the track could not be separated from the operation of traffic. No matter how simple it might seem for the state to put the construction of the lines into practice, operating them once they were built appeared arduous and contrary to the nature of things. Massed against the Department were the ranks of the liberal economists who had a wide hearing in parliamentary circles and favored an "English" solution to the problems of the railways. According to these authors, railways should be constructed and operated solely with private capital, which would only be committed if they proved effectively remunerative. To quote one writer,[4] it was advisable "to let the capitalist instinct seek out the most advantageous investments for itself". The liberal stance was therefore categorically opposite to that of the Department.

Alexis Legrand, the Director and spokesman of the Department, justified state control of the construction and operation of at least the main lines by arguing against the liberal viewpoint. Railway lines in France would be very costly because they would be very long. The demands of such an undertaking outweighed the capacities of private industry. Only the State could build the railways because it alone sought no return on its capital. "Interest", said Legrand, "is returned indirectly a thousandfold, by the prosperity of the country, by the increased value of the land, by the progress of trade and industry."[5] But Legrand was over-optimistic as regards the funding capacities of the State. The resources needed to set up infrastructures and to purchase equipment were indeed lacking. The State was already committed, with no possibility of retracting, to a vast scheme to renovate the road network and to build more navigable waterways. There were no funds that would enable the Department to carry through an extreme solution. Again, "private interests" themselves could not be mobilized properly without some measure of concertation and cooperation with the State. A compromise would have to be reached. But what private interests were in fact ready and available?

In Great Britain, the first railway lines were built by industrial and merchant groups anxious to create better conditions for the carriage of their products[6]. It was industrial capital that largely engendered railway capital up to at least 1850. In France, it had seemed for a time as if a similar process might be sketched out: the lines of the Loire network, and the Talabot network round the Grand Combe are examples that spring to mind. Above all, there was the Alsatian network created by the industrialists of Mulhouse.

All the same, it very soon became clear that industrial capital was sparse and local commercial capital too dispersed to meet the needs of this type of investment. Under these circumstances only a decisive commitment from the great bankers would allow the challenge to be taken up. For the banks alone could create the confidence required to draw the mass of available capital towards the new enterprises. That commitment was slow in coming: it was timorous in the 1830s and did not become wholesale until after 1844. Until the 1840s the Parisian banks had directed their power and efforts towards the financing of international trade and the investment of French and foreign government money. Tendering for railway lines signified an abrupt change in activities and could not be contemplated without prior reflection and study.

Furthermore, the hesitation was legitimate. Alexis Legrand recognized this himself in 1838. "With regard to the railways, everything is uncertain, everything eludes prediction; it is impossible to assign a destiny to these new enterprises."[7] There was no reasonable basis for evaluating building costs or operating costs or profits. Even the technical possibilities of the system were open to question. Originally, railway technology had seemed doomed to failure. The experiences of the first three lines in France, built in the mining basin of Saint-Etienne from 1827 to 1834 (Andrézieux - Saint-Etienne, Saint-Etienne - Lyon, Andrézieux -

Roanne) were very disappointing. A report drawn up in 1836 concluded that these railways were "industrial misconstructions".[8] There was nothing to prove that the technology could be transposed to the level of a national network. Heading the skeptics was the scholar François Arago (1786 - 1853) who, in a speech to the House in 1838, referred to "the imperfect state of the art". He saw the locomotive as a totally unfit and unfinished machine and went on to illustrate the inferiority of railways compared with canals for the carriage of freight and above all expressed doubts as to the "pecuniary productivity of the railways".[9]

These hazards explain the waverings of the actors who found themselves engaged in a game with unknown rules. It is important to summarize these complex developments full of false starts and all kinds of disillusionments before attempting to assess the extent of those hazards and describing the way they were gradually attenuated.

Hesitation and failure

In 1832 an agreement approving the idea of a concession system was reached fairly quickly. But for the next twenty years the solutions adopted were as many and different as the concessions granted, despite the general principles laid down under the law of 1842. The central issue concerned the duration of the concessions. This depended on two parameters: firstly, on the amount of the investment (the "initial outlay"), which was calculated according to the costs of the works and the way those costs were shared between the State and the companies, and secondly, on the value of the annual income.

The solutions adopted during the July Monarchy failed. The Department, almost as a matter of course, refused to grant concessions long enough to ensure a satisfactory return on capital. Moreover the costs of the works always exceeded estimates. The system set up between 1837 and 1846 was in a state of crisis by 1847 and collapsed altogether after the 1848 revolution. In 1852, Napoleon III succeeded in laying the foundations of a new system by turning to account many of the elements of the earlier methods and combining them in a homogeneous fashion.

Between 1830 and 1852, the length of the "public interest" lines increased from 148 kilometers to 7,400 kilometers and the length of the lines in operation from 38 kilometers to 3,870 kilometers. When

Napoleon III came to power there were still barely 4,000 kilometers of railway lines in France. The leap forward was taken in the 1850s: by 1860, 9,500 kilometers of lines had been built and almost 17,000 granted. The initial outlay or rather the whole of the construction costs incurred from the outset attained 986 million francs in 1847, 1,450 million in 1851 and 4,725 million in 1860. Receipts from high-speed and low-speed transportation and traffic increased accordingly (see Table 1).

Table 1: Evolution of Receipts and Traffic from 1841 to 1860

	Receipts		Traffic	
	High Speed	Low Speed	Passenger	Freight
	million francs		thousand units per kilometer	
1841	7.9	4.6	0.112	0.037
1846	24.9	14.1	0.327	0.126
1852	76.7	52.3	0.988	0.621
1860	176.0	228.5	2.52	3.14

Up to 1854 the receipts from high-speed transportation and passenger traffic evaluated in units per kilometer were greater than the receipts from low-speed transportation and freight traffic. Unlike Great Britain, France had constructed its railways with a view to catering for passenger traffic rather than the carriage of freight. The rapid growth of freight traffic came as a surprise but after 1850 was regarded as a major target of commercial operations on the networks. The skepticism displayed by Arago and so many others played a significant role in the repeated failures of the 1830s, those of the many private lines and also that of the global solution proposed by the State in 1838. To list all these in detail would be tedious. The chief failures concerned the Paris - Rouen line in 1835 and the lines from Paris to Belgium, Lyon to Marseille and Paris to Chartres in 1837.

In 1838, the Department of Ponts et Chaussées nonetheless decided to try its luck once more. The Minister presented a project for the classification of four main lines (Nord towards Belgium, Ouest towards

Le Havre, Sud-Ouest towards Bordeaux, Sud-Est towards Marseille) and for immediate construction by the State of the Paris - Lille, Paris - Orléans, Paris - Rouen and Marseille - Avignon lines. The ensuing debate, dominated by Arago's speech, ended in defeat for the government. "The vote", wrote Alfred Picard, "disqualified the State not only from operating but also from building the railways."[10] In his speech, Arago, deriving inspiration from liberal thinking, had vaunted private enterprise. And indeed concessions to private companies augmented after 1837. The first and founding model was the line from Paris to Saint-Germain.[11] Emile Pereire, the promoter, succeeded in convincing one banker, Alphonse d'Eichtal, and Eichtal was able to exert his influence on others including the Rothschilds. In June 1835 parliament agreed to grant the line directly to the company formed by Emile Pereire. His "cahier des charges" drew up a series of stringent technical constraints, all of which made up the central features of the future national network. The concession was declared for 99 years and operations were placed under the control and supervision of the Department. From the very start, railway development was contained within the bounds of rigid administrative regulations.

After 1835, the Paris - Saint-Germain concession was followed by the Montpellier - Sète and the two lines linking Paris to Versailles, one along the right bank of the Seine and one along the left bank. By 1838 the euphoria was general: in January there was the Strasbourg - Basle and in May a Paris - Rouen and a Paris - Orléans. But it was not to last: in 1839 the underwriters of the securities of the new companies proved reluctant to continue their payments. The Paris - Rouen concession was terminated. As Dufaure[12], the rapporteur of the 1842 law, put it: "The biggest fortunes backed out". What was more, the estimates for the Paris - Versailles lines had been exceeded threefold. A loan had to be obtained for the Versailles - Rive Gauche and the Paris - Orléans, where initial estimates again had been left far behind, had to have recourse to the interest guarantee. Such methods of assistance, however, were strongly and rightly criticized by a whole section of public opinion, especially liberal opinion. Even in 1837, the deputy Duchâtel had shown that these schemes amounted to making "the concessionaires lose interest in good management". In 1839, Lamartine defined this solution as follows: "C'est le malheureux contribuable, constitué par la loi, le croupier de l'agioteur". (The unlucky tax-payer has been made the lawful backer of the gambler.[13]) In the 1850s, the interest guarantee was to reappear nevertheless.

The Department had been looking for a way of taking advantage of the difficulties afflicting private investors in order to make up for the ground lost in 1838. It had by no means abandoned the idea of having the main-line railways built by the State. In 1840, work began on two sections: the line from Lille to Valenciennes on the Nord line and the line from Montpellier to Nîmes on the Sud-Est. In 1842, the government's main concern was to show without any shadow of doubt that it was prepared to act in spite of the refusal of 1838 and it accordingly put forward a general law as a basis for defining a program. Pessimism was still the keynote: "Our wealth is moderate", observed Dufaure. "Our foreign trade cannot bring in the admirable resources for domestic improvement which the foreign trade of Great Britain supplies."[14] Costs, therefore, had to be shared: the State would build the infrastructures only, which would account for about half of the total expenditure, and it would retain ownership of the track. The companies would build the superstructures and would operate the lines on a lease basis, i.e. over a very short period of time. This system was applied only to a very limited extent.

The loss of confidence in the future of the railways proved shortlived. Not only had the State made it quite clear that it was determined to go ahead by opening several building sites; its concession of the Paris - Saint-Germain had also proved a success. It had been an experimental line and the experiment had worked. The financial results were perhaps not quite as brilliant as might have been hoped but were very respectable even so. The railways could at last be regarded as an "industry". This was doubly confirmed in 1843 by the results of the Paris - Orléans and the Paris - Rouen. The new Paris - Rouen had been granted in 1840 to a company disposing of capital that was partly of English origin. Its first 145 days of operation from 9 May to 30 July 1943 were an outright triumph and for the year its receipts rose to a total of 4.4 million francs against an outlay of 2 million francs. The novelty lay in the freight sector, which had boomed with the cheaper rates beyond all expectations. The results of the Paris - Orléans were equally remarkable.

Enthusiasm flared up once more and concessions went on being granted until 1846. But the Department used this as an argument to be more demanding. The "cahiers des charges" became more and more rigorous, furthermore the Department attempted to play on the competition between banks with a view to shortening the length of the concessions. By then concessions were by adjudication and were no longer granted directly. From 1842 to 1848, 4,000 kilometers of lines were allotted in this way for a total capital of one billion francs, and the average length of the concessions dropped to 46 years. Only four lines were managed under the work-sharing scheme provided by the law of 1842. All the other lines were entirely built by the companies. The decisive factor was the massive commitment of the great banks: the Rothschilds held twelve managerial posts on eight boards. Laffitte held nine posts on six boards and the Hotinguers sat on three boards. The objective pursued by the banks was twofold: 1. to control the issue of the securities required to fund the railways; 2. to guarantee profits which it was hoped would be substantial, for the shareholders.

The boom on the stock exchange was unprecedented and far more wide-ranging than that of the 1830s. But the crisis broke out in the autumn of 1846 when shares plunged abruptly, largely because of the railways. This slump, according to a financial journalist R. Bloch, was due more to the fear of what might happen in the future than to difficulties actually felt at the time.[15] Yet the drop was warranted: the forecasts on which the entire economy of the agreements signed between the State and the companies rested had proved false. The cost of the works had been underestimated. In 1842, the average outlay expected for building was 300,000 francs per kilometer. The actual outlay which appeared in 1847 was in the region of 500,000 francs. Traffic density admittedly had come up to the mark. It was not activity that was lacking. Operating costs had simply been far greater than planned and the crisis laid bare the full complexity of the technical details involved in managing a network. The miscalculations were technological not economic.

Faced with these difficulties and miscalculations, the companies made two requests of the government: an extension of the duration of the concessions in proportion to the amounts overdrawn, and a lowering of customs duties on railway materials in order to reduce construction costs. Both were rejected although a few allowances were made on the first point. The quotations for railway shares crashed in 1847: Saint-Germain went from 800 to 340 francs. When the 1848 revolution broke out, the friction between the State and the companies was acute. On 24 February the provisional government launched a program to buy up the shares but the failure of the revolutionary days of June put an end to the talks. The dispute remained intact, and the Department of Ponts et Chaussées tried to take advantage of the political situation to consolidate its assets. The construction of the Paris - Lyon was continued. But no sound area of agreement could be reached with the companies. While a violent opinion campaign, with anti-Semitic overtones, raged against the companies, the government refused to take global measures and restricted its action to partial adjustments. This is why railway investment never recovered its impetus under the Second Republic. After seizing power on 2 December 1851, the future Emperor Napoleon III gave the needed and decisive thrust which sparked off reinvestment in the railways. A 99 year concession was granted to the existing companies and to the new companies created after 1852 by grouping old and new holdings. This meant that the charges on loans could be spread over a far longer period and would be greatly eased in the immediate future. Consequently, to quote the directors of the Nord company, it ushered in an era of "handsome dividends". The extension also enabled the companies to adopt a policy of lower rates on a large scale.

Between 1852 and 1859, the aforementioned organization into networks was completed. "Our six main networks are now created", declared the Minister of Public Works in 1876 at the funeral of Franqueville, the man who had been the chief architect of the conventions of 1859. The central organization was powerfully constituted. It could be left to itself. It was sure to survive and revenues would suffice not only to meet operating costs but also to guarantee ample return on the capital committed in the vast undertaking.[16] The economy of the system rested on the monopoly and the interest guarantee, or recourse to public funds to service any deficit. The partnership between private interest and the State had therefore been achieved on bases totally contrary to liberal principles. The companies were obliged to maintain railway lines that would presumably run at a loss and in return the State granted them a monopoly and a guarantee which, admittedly, had to be refunded. In theory, the State acted merely as a relay; present deficits would be financed by future surpluses. In practice, however, the State demanded something in return. The companies lost the control of rates. They were prohibited from signing private agreements with senders. They were subjected to financial control and they had to obey the rules of administrative accounting. The price paid

French Railway Network in 1855

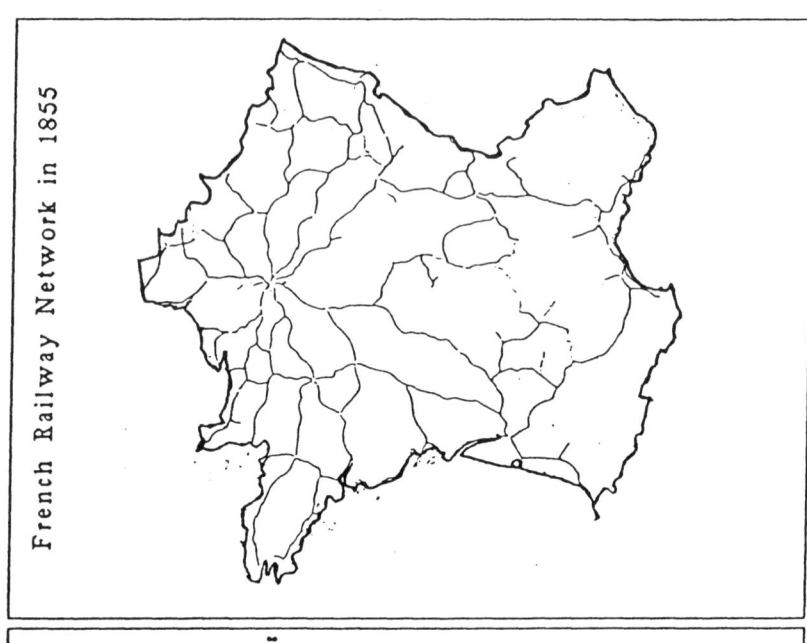

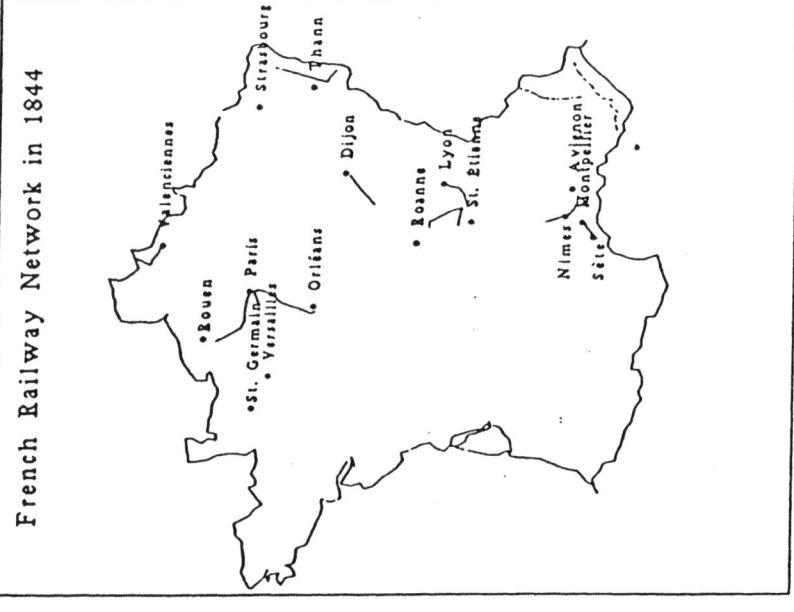

French Railway Network in 1844

Railways in France 83

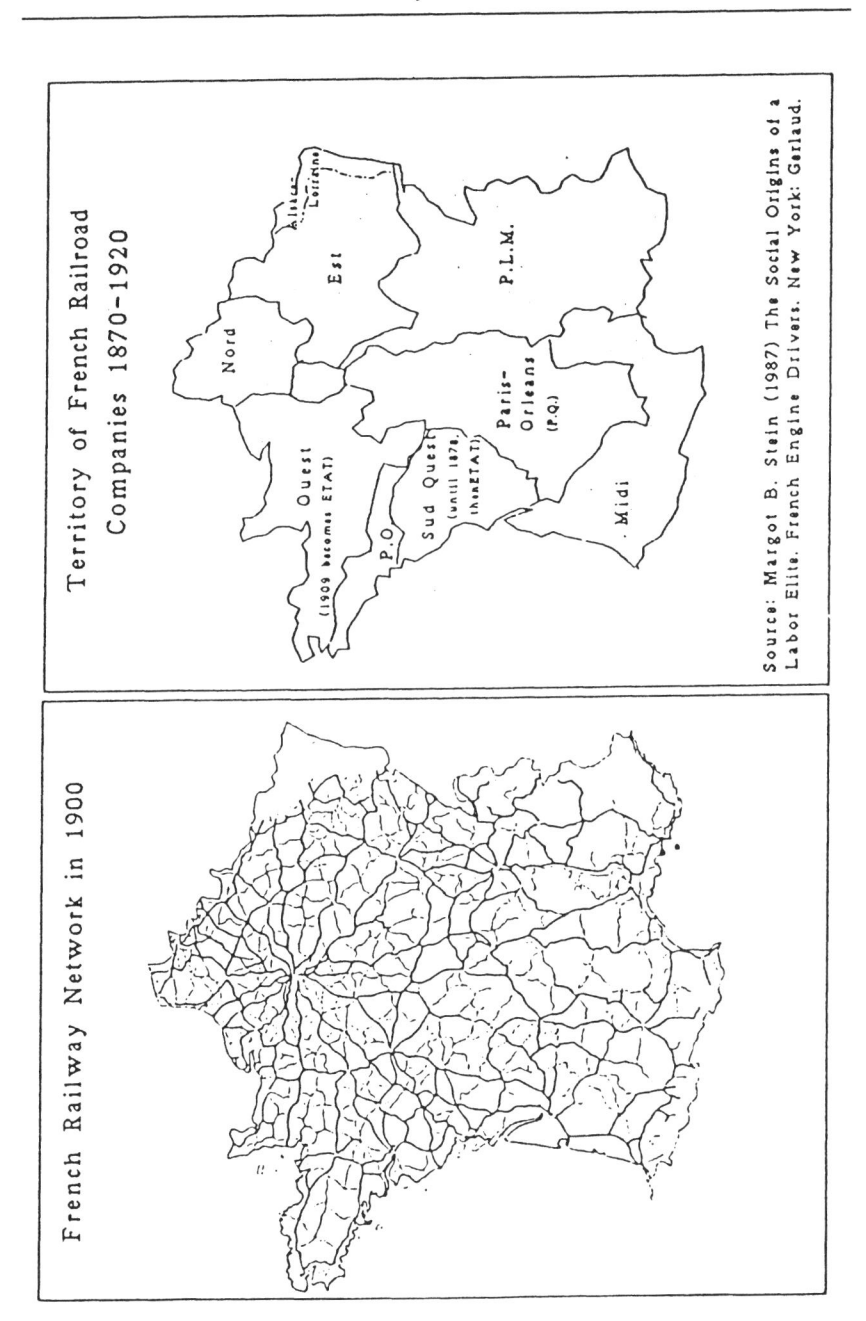

Territory of French Railroad Companies 1870–1920

Source: Margot B. Stein (1987) The Social Origins of a Labor Elite. French Engine Drivers. New York: Garland.

French Railway Network in 1900

for the monopoly, the extension of the network and the investment security was administrative tutelage and banker's control on the issue of bonds.

The emergence of the technical system: 1832 to 1859

As the institutional system moved on to an even keel, the initial stage in the evolution of the technical systems was completed. Operations became less hazardous and more efficient. With the extension of the concessions and the introduction of technology better adapted to practical requirements, a long-term commercial strategy could be adopted. The uncertainty hanging over the future of the railways, which had reappeared with the 1847 crisis, was swiftly dispelled. Therefore the real point of departure for the expansion of the system may be said to date from the 1850s.

The construction and operation of the railways had mobilized all existing techniques from the very start. Yet none of these had fully answered initial needs and seemed quite inappropriate for the new needs which were constantly appearing. The adaptation of existing techniques to operational demands was only achieved after a long process of apprenticeship. This is why Arago's standpoint was wrong: in order to develop technology it was essential to plunge into the thick of the fray despite "the imperfect state of the art". For creation lay in experimentation. The technical system of the railways in fact comprised three lines of development: the use of energy, the use of materials, and the use of long-distance communication.

The French locomotive industry became established between 1837 and 1847. During those years, 102 locomotives were built in France and a further 102 were imported; from 1842 to 1847 the figures rose to 384 French engines and 60 foreign engines. By 1854 France was producing 500 locomotives a year. This marked its peak. It was a formative period in a new sector of activity which, following the classic model, witnessed considerable mobility among firms. François Crouzet[17] listed 17 registrations between 1834 and 1861 but observed that only six firms actually survived and were of real importance. Most of the manufacturers had previously been mechanical engineering firms. Yet it has to be admitted that until at least the early 1850s the locomotive remained an engine awkward to use. It was only in

1849, with the introduction of the Crampton locomotive, manufactured in France by Cail, that the problem of speed could be solved. The Crampton continued to be used for the majority of express trains into the 1880s although its shortcomings soon became apparent. It was very slow in starting and proved dangerously stiff on curves but, above all, while it had to haul increasingly heavy loads its power could not be augmented. In the early 1860s the Nord and the P.L.M. companies began to look for a substitute. Most of the French locomotives employed for the carriage of freight were derived from the Stephenson model with three driving-axles and variable expansion. Between 1846 and 1849 the factory at Le Creusot brought out a series of modes of this type, which remained in service until the 1880s. This engine, however, was totally inadequate for the heavy trains weighing 450 tons which the Nord network attempted to inaugurate in 1852. The company consequently chose a locomotive based on the engine designed by the Austrian engineer Engerth in 1851 to cross the Semmering pass on the Vienna - Trieste line. This order heralded a generation of robust and powerful locomotives, perfected after many modifications, which were to hold sway over French networks for 50 years.

Hitherto, no truly satisfactory solution had been found for climbing sharp gradients. The escapade of the "atmospheric railway" tried out on the Saint-Germain line in the 1840s was a sign that some engineers thought the steam locomotive had exhausted its possibilities. No sooner was the atmospheric railway built in 1846, however, than it proved inefficient; in 1859 its operation was stopped after a serious accident. The case illustrates well the technological insecurity that prevailed in the 1840s. It was not until the 1850s, as said before, that the power, speed and haulage capacity of locomotives attained standards high enough to establish the new techniques once and for all. During that decade research was primarily directed towards fuel economy and was spectacularly successful largely owing to the adaptation of the bars of the grate. This meant that very low grade coal could be burned instead of coke and in much smaller quantities. The engines manufactured over those years were kept in use on the networks until well into the 1880s, when a second technical stage was passed.

The improvements in locomotives as well as in rolling-stock and tracks demanded fundamental modifications in the materials employed. The metallurgical industry could cope with the needs of the railways neither qualitatively nor quantitatively. There is ample evidence of this, referring to rails, axles, iron tyres and points. The accident which occurred on 8 May 1842 at Meudon on the Versailles - Rive Gauche line deeply impressed public opinion. In all there were 44 victims, who died under appalling conditions.[18] The accident had been caused by a severed axle, perhaps due to a broken spring. Another accident, which caused 14 deaths, occurred in Fampoux near Arras on 8 July 1846. Although it was attributed to a landslide its seriousness was undoubtedly due to faulty equipment. Poor quality materials not only entailed a permanent risk of accidents, they rendered operating costs unpredictable. "The companies", noted a receptionist belonging to the Nord network, in 1857, "found themselves liable for very high costs for rails that were taken out of service".[19] The renewal of rails and iron tyres soon became a steady drain on finances. Both items were fragile and both rapidly showed unexpected signs of wear and tear.

The companies reacted by obliging their suppliers to agree to rigorous manufacturing controls and very long guarantees. In the 1850s, guarantees for the Nord and Est networks were for three years. Deliveries were tested carefully and deficient lots were systematically rejected. Suppliers were consequently put through a ruthless selection process. The quality of the rails improved and most of the companies completely overhauled their lines using the new materials in the course of the 1850's. An equilibrium appeared to have been reached. But it was not to last.

Railway operations had been in need of an efficient system of long-distance communication from the very start, if only for safety reasons. It was essential to avoid collisions of succeeding trains and at junctions. Safety on the open track had originally been ensured by itinerant watchmen or guards at permanent posts, while sensitive points (junctions, stations, level crossings) were fitted with mechanical devices that could be worked from increasingly long distances. But these transmission systems became more and more vulnerable. During the day, the indications given by the pointsmen, gate-keepers, and permanent-way brigades were deemed adequate to provide the engine drivers with accurate information on the movement of the trains and any occurrences on the track. At night, itinerant guards, who covered four or five kilometers, were used. Besides there were precise regulations concerning the running and timing of the trains. The most widely adopted solution was to allow an interval of 10 minutes between departures.

None of these arrangements proved satisfactory and the State exerted strong pressure on the companies to urge them to find a more reliable system of protection. In 1847 a special Commission was appointed to investigate the matter. In its report it concluded that the railway was a far less dangerous means of transportation than earlier modes. It endeavoured to clarify signalling instructions and to ensure a minimum of harmonization among the companies. The companies were obliged to increase their permanent posts on the route where traffic was heaviest. Furthermore, a number of important innovations were being made. In 1845 electric telegraphy was tried out on the Saint-Germain. The introduction of cables along the lines, connecting the stations, became almost general in railway building after 1846, although the use of the system was not exclusively reserved for the railways. In the 1850s telegraphy became "the essential complement to the art of operating the railways".[20] It rendered incalculable services in regulating the running of trains and made for greater safety. Yet it provided only a very partial answer to most difficulties. There was still no guarantee of safety on the track or at junctions, or even of optimum use of the installation. In the 1850s the first signalling devices based on automation processes made their appearance. Already in 1847, Regnault, an engineer of the Ouest network, proposed a telegraph system to link up the different permanent posts to make the application of the principle of the block system feasible. As a result the running of trains would be based on distance rather than time. But the application of this method remained limited because it depended on the accuracy of the operators who performed the manoeuvres. As there was no means of connecting the movements of distant signals automatically only very partial and inadequate use of the block system was possible.

It was when the technical system was nearing its initial stage of completion in the 1850s that the administrative organization of the networks became established. The first model, evolved by the Paris - Saint-Germain, was retained though in an improved and more harmonious version. The pattern, broadly speaking, consisted of a division into specialized technical branches and a strongly hierarchized structure within each branch. Alfred Picard noted that the difference between one company and another were negligible because "the general needs were the same, which led naturally to similar if not identical solutions".[21]

At the head of the company, appointed by the board, was either a committee (the Nord network had a committee of five members) or a director, whose role was statutory only on the Paris - Orléans. The director was responsible for four divisions: the central division, the division of operations, the division of traction and materials and the division of track and buildings (travaux). The central division was originally a secretariat with a few additional services such as a pension scheme, a legal department and a medical centre. In 1856, the number of staff in the central division amounted to only 2% of the total staff employed by the companies. In fact the division of operations dominated all the others. It was made up of two sections: a very powerful central section, which defined commercial policy, organized the running of the trains, accounted for expenditure and controlled receipts, and dealt with claims and disputes. It in fact defined the general policy of the enterprise and enforced its implementation. The second operational section was decentralized and the networks were divided into areas placed under inspectors who had authority over the officers in the stations and on the trains and maintained permanent relations with customers.

The traction and materials division also had a central section, which was chiefly responsible for the management of equipment and for ensuring appropriate attention to orders. The equipment section proper dealt with the preparation, maintenance and supervision of those orders. The traction department was organized into a number of depôts. The staff employed by the division as a whole represented 31% of the total staff in 1856. All were highly skilled workers or engine drivers.[22]

The track and building division employed the remaining 30% of the total staff, who were far less skilled and remunerated. Again there was a central department as well as local branches in charge of the routine and often thankless tasks involved in construction and maintenance. The rail network was divided into districts and these were sub-divided into sectors.

While each division was governed by its own central department, which defined general principles, controlled the functioning of the system, carried out practical research and dealt with orders, the task of administering the different sections and services as a whole remained incumbent on the central department of the first division. This department was in charge of accounts and was generally required to define the strategy of the work to be undertaken. It evolved commercial policy and also the broad lines of investment policy. The system was a coherent one. It subordinated the duties of the two "technical" divisions (traction and materials; track and buildings) to the division which was in direct contact with the customers and therefore was in the best position to assess needs. The efficiency of the system owed a great deal to this subordination. At the same time, however, the director of operations on the Nord network was closely supervised by the board and on the other networks by the director general, both of whom unfailingly recalled the need to bear in mind monetary constraints. Company management was in fact a compromise between the demands of the divisions and the financial preoccupations of the board.

The second general characteristic of the organization of the railways was the strength of the structural hierarchy. The differentiation of duties rested on a complex scale of qualifications linked to the technical nature of the task. At the top of the hierarchy was the engine driver and at the lowest level the track-watchman. The scale of duties obviously corresponded to a scale of wages and of prestige. It was completed by an increasingly strict promotion scale and obeyed increasingly formal rules. But the other criterion of differentiation in the hierarchy of duties and grades concerned authority. The proper functioning of the system, particularly where safety was concerned, depended, in the eyes of all the directors, on staff *obedience*. That obedience rested on the integration of the staff in a system of very severely hierarchized powers, on the respect of a considerable corpus of regulations and, lastly, on a system of rewards and sanctions which an administrative controller justified in 1882 as follows: "The active service of the railways makes harsh demands".[23] The entire organization in fact rested on the technical constraints of the system: the separation into divisions reflected a concrete division of tasks, the hierarchy reflected differences of skills but also the need for rigorous authoritarianism in order to limit the risk of accident or incident. It is fair to say that this system, which was at once multi-purpose and centralized, hierarchized and authoritarian, worked efficiently throughout the greater part of the nineteenth century and permitted the adaptation of the network to pressure of traffic, diversification of customers' needs and technological change.

2 The expansion of the system

The extension of the network

From 1852 to 1914 the history of the railways in France, as in the other European countries, was marked by a general process of expansion. Line kilometers increased and investments in the old lines climbed in a manner that was more and more difficult to control while traffic developed at a steady rate under the effects of a no less steady drop in rates. This evolution is summarized in table 2. The dates have been chosen to show the principal turning points and are not arbitrary.

Table 2: Growth of the Network

Year	Length Operated Thousand km	Passenger Traffic P.K. Billion	Receipts Million francs	Freight Traffic T.K. Billion	Receipts Million francs	Average Product of P.K.	T.K. centimes
1852	3.6	0.99	66	0.6	63	6.67	0.1
1859	8.8	2.70	139	2.8	244	5.14	8.8
1873	18.0	4.33	230	8.1	560	5.13	6.9
1882	25.4	6.73	328	10.7	744	4.87	6.9
1890	32.8	7.90	352	11.6	753	4.45	6.5
1896	35.5	11.10	423	12.9	830	3.81	6.4
1906	38.4	14.70	536	18.2	1070	3.64	5.8
1913	39.5	19.30	663	25.2	1350	3.43	5.3

P.K. = Passenger Kilometers
T.K. = Ton Kilometers

The examination of the causes and modalities of the extension of the network will be followed by a description of the forms and stages of traffic growth and the way they are related to the growth of productivity and the drop in transport prices.

The extension of the network was not due to the companies' desire to increase the line kilometers. On the contrary, it was the result of an "all-consuming thirst for railways", to quote Louis Girard. Each canton wanted its station. Yet it was not without circumspection that the companies agreed, in the 1860s, to the construction of new lines in order to add a "third network" to the initial "old network" and to the "new network" of the 1850s.

After 1865, what can only be described as open warfare broke out between the "big companies" and small companies formed by local dignitaries round a handful of speculators. As these companies were not eligible for state concessions, they availed themselves of the facilities provided by a law on local lines voted in 1865. This law allowed departmental authorities (prefects and councils) the right to grant such lines directly. It seemed a simple matter to connect one department to the next and to create networks that would be in a position to compete with the main lines. The speculation failed: from 1875 onwards, the companies went bankrupt one after the other. However, the lines had been built. Some were integrated into the main system, others were bought up by the State. At the same time, in 1879, urged on by Freycinet[24], the Minister of Public Works in the first truly Republican government, the State embarked on a vast construction program designed to cover 17,000 kilometers, soon to be known as the "electoral railways". The State undertook the building itself but the financial crisis of 1882 brought the venture to an end. The 1883 conventions settled all the difficulties which had arisen from the chain of contradictory policies by integrating the lines built by the small competing companies and those of the Freycinet plan into the existing main networks (Ouest, Nord, Est, P.L.M., Midi, P.O.). The Etat network was created between the Paris - Orléans and the Ouest and placed under special management. In 1908 it absorbed the Ouest network after taking over the company. The construction of the lines contemplated by the Freycinet plan was staggered and, while remaining the responsibility of the State, was to be completed by the companies themselves.

Thus bonds underwritten by the State proved to be the chief financing instruments for railway investments in France. In 1882, of the 12.2 billion francs spent since the beginning, 26.4% were supplied by the State, 16% by company shares and 57.6% by the issue of bonds. Bonds accounted for 78.2% of the companies' capital then and 91% in

1913. But the nature of the investment had altered considerably. Even in the 1850s it was no longer devoted solely to building costs. A series of gross investments has been reconstructed here (see Table 3) by adding together the expenditure involved in the "initial outlay" given in the official statistics and the expenditure incurred for the renewal of equipment and heavy maintenance of plant evaluated on the basis of the companies' accounts. Equipment and maintenance costs represented 35% of the total expenditure from 1875 to 1884 and 37% from 1885 to 1913. Consequently almost two-thirds of investment costs consisted

Table 3: Investment Expenditure and Productivity

	Investment Expenditure in Current Francs (Million)		Factor Productivity Index Base 100 = 1913
	Gross Investment	New Investment	
1845-49	163.0	144.3	
1850-54	180.8	161.8	44.7
1855-59	406.4	354.9	55.2
1860-64	396.7	321.7	60.6
1865-69	317.8	225.7	67.2
1870-74	290.9	183.2	66.4
1875-79	380.8	224.9	66.5
1880-84	578.1	396.0	60.5
1885-89	423.2	275.3	60.5
1890-94	401.1	241.0	65.4
1895-99	375.7	200.0	73.1
1900-04	531.9	342.0	80.2
1905-09	562.0	341.0	86.2
1910-13	693.7	476.6	95.6

Source: F. Caron, Histoire économique de la France. 19.-20. Siècles. Paris: Colin, 1984.

of new investments. Until the 1880s construction represented about three-quarters of these amounts, by 1906 one half, and by 1913 one third. The remaining was made up of "supplementary" costs incurred for the extension of plant and equipment on existing lines with a view to meeting fresh needs arising from the increase in traffic. The "initial outlay" for the railways was an account that could never be closed.

Growth of traffic

The higher supplementary costs were directly due to the pressure of traffic. Up to 1873 the carriage of freight increased far more quickly than passenger traffic, multiplying by 14, whereas passenger transportation only multiplied 4.3 times. At that time passenger receipts accounted for only 41% of the receipts from the carriage of freight. But from 1873 to 1896 the evolution was reversed: passenger traffic augmented at a far faster rate, multiplying by 2.6 against 1.5 for freight. In 1896 passenger receipts had grown to half of the receipts for freight. From 1896 to 1913 growth in both sectors was almost even with freight slightly in the lead (+ 95% instead of + 75%).

The differences in rhythm during the three periods corresponded to the dynamics of the general economy but also matched and complemented the evolution of transport prices and productivity. Until the early 1870s, in fact, traffic and receipts exploded and freight rates dropped steeply, as did passenger rates although not quite to the same extent. The reduction in freight rates was largely achieved by taking into account the differential cost prices for transportation associated with an active commercial policy. The director of operations for the Nord company, Jules Petiet, anxious to justify the adoption of reduced rates for coal carried by special trains as well as the purchase of Engerth locomotives to haul them, declared to the board in 1855: "If coal is to be transported in large quantities there must be a reduction in rates even in spots where there is no competition from canals".[25]

The said commercial policy systematically turned to account the opportunities the technical system of the railways had to offer. It exploited both its capacity reserves and productivity gains from the innovations introduced. A global factor productivity index (capital, work, energy) has been calculated for the French railways as a whole (see Table 3). The index rose at a rate of 2.06% a year from 1851 to 1873, principally owing to the aforementioned capacity effect and the progress in steam traction.

Conversely, the period from 1873 to 1896 was characterized by a long-term stagnation in productivity. Basically this can be explained by the extension of the network: the majority of the new lines had too little traffic to permit optimum use of the inputs. At the same time, on the main lines built earlier, operational difficulties occurred in the early 1870s and again in the early 1880s through the bottle-necks caused by sporadic and abrupt increases in traffic. In the 1880s the situation was reversed yet again: freight traffic stagnated owing to the economic crisis and the network experienced a general over-capacity. The productivity index fell to its lowest level between 1887 and 1888. But throughout that decade passenger traffic continued to increase (+17.4% from 1882 to 1890) and passenger rates continued to drop. From 1890 to 1896 the growth in freight traffic, compared with that of passenger traffic remained very weak (11% versus 40%). It was the dynamism of the "passenger" service which compensated for the virtual stagnation in freight traffic from 1882 to 1896. The original disregard with which passengers travelling in the cheaper second and third classes had been treated was no longer accepted.

It was likewise in the 1880s and above all the first years of the 1890s that more and more measures were introduced to bring out the mass aspect of travelling. In 1891, the director of operations of the Nord network put forward the very argument used by Petiet for his coal trains in 1855 in order to justify a sharp reduction in the price of return fares for tourists. "Our aim is to reach those many customers who will be incited to travel if sufficiently low prices are offered for a suitably long journey".[26] To a very great extent this policy was the direct consequence of the 1883 conventions. The companies had formally undertaken to develop passenger traffic just as they had promised to improve travelling conditions, even in the cheaper classes.

The financial forecasts which had warranted the 1883 conventions proved unfounded: the authors had hoped that the surpluses of the old lines would cover the deficits of the new ones but this was not to be. The stagnation in traffic caused a severe slowdown in the growth of receipts while expenses continued to rise sharply owing to the extension of the network and the new charges laid on the companies under the conventions. The interest guarantee for the main lines rose from 43 million francs in 1884 to 101 million francs in 1893. In actual fact a rift was opening between the networks which often ran at a deficit and had to resort to the interest guarantee and the networks with surpluses like the Nord and the Est.

From 1896 to 1913 freight traffic increased at a slightly faster pace than passenger traffic (95% instead of 75%) and rates dropped again steeply. A rise in productivity at a rate of 2.3% a year was achieved although construction continued. This was due to the systematic application of the innovations made during the previous period and to improved modes of operation. But this technical achievement was not matched by an equivalent financial success. There was no resuscitation of the "handsome dividends" of the Second Empire. The higher wages and shorter working hours demanded by the State together with the rise in the price of supplies increased costs out of all proportion. Moreover transport prices could not be raised. Looking for higher profits had ceased to be the chief target of railway operations. The companies attempted to regularize their dividends by attenuating abrupt fluctuations in the net product. In any event, the gains made possible by the rise in productivity served to reduce the men's working hours, to service the deficits of the unprofitable lines and to lower tariffs. Railway profits from then on were thoroughly socialized.

In spite of the number of unprofitable lines and the strong pressure of traffic, the system succeeded in adapting as it became socialized. Henceforth the companies acted as the managers of public services although they preserved their initial spirit of private enterprise. Despite the weight of state tutelage each company enjoyed considerable autonomy in decision-making and maintained its own technical philosophy. In view of this it may seem an exaggeration to offer a global analysis of the French technical system, but the many identities and resemblances among the companies nevertheless justify this approach.

Technological adaptation

The years from 1873 to 1896 must be regarded as years of gestation which produced a new technical system totally different from that of the 1850s. Each of the three branches discussed above evolved at its own pace, on the whole the opposite of that noted in the previous period. Spectacular strides were made in locomotive techniques from 1842 to 1873 but progress in the field of materials, let alone in transmission processes, were far less brilliant. A distortion consequently appeared. The technical system had lost coherence: the power of the traction engines aggravated the wear and tear of the rails and made the regulation of traffic more dangerous. The technological mutation which took place between the 1870s and 1900 must therefore be understood first of all in terms of the harmonization of the different branches. This was achieved in the 1860s and 1870s through the pressure of traffic. This growth created "bottle-necks" and they increased congestion and accidents. Accidents were particularly numerous between 1871 and 1873 and again between 1878 and 1882, i.e. during the two periods when traffic augmented to an exceptional degree. Furthermore the irregularity of the pressure made it difficult, within the framework of an outmoded technical system, to control costs. The economies of scale became negative. But in the 1880s and 1890s another constraint made itself felt, that of social demand. Freight traffic as seen before was experiencing virtual stagnation by then, although passenger traffic maintained its growth, while the State and parliamentary opinion pressed for reductions in rates and improved services as regards speed and comfort.

The materials sector had been upset by developments in the use of steel. Already in the 1860s, the railway companies had favored the adoption of the Bessemer method for the manufacture of rails. The first orders dated from 1863. The maintenance of the parts of the track which did heaviest service in fact increasingly involved considerable expenditure. In 1872, steel prices were such that the replacement of iron became worthwhile for the majority of the networks, particularly as steel was over ten times as long-lasting. In 1879, the authors of the "Statistics for the Mineral Industry" observed that rails made up "the greater part of steel production by the new methods". Railway orders were to play as great a role in the development of the Thomas Gilchrist method between 1879 and 1882 as they had in that of the Bessemer method in the 1860s and 1870s. The influence of railway needs on the technologies of materials was by no means restricted to rails and ordinary steel however. The technology of metal bridges was also highly perfected in France, as was that of alloy steels employed for the more sensitive parts of the tracks and equipment.

The French railway companies likewise played an important role in promoting the use of electricity. Electricity was introduced very precociously at the beginning of the 1870s to light large areas: stations, marshalling yards, depôts. Electric light greatly increased the productivity of both the machines and the men and helped to reduce the risk of accidents to a very great extent. The research undertaken during the 1880s and 1890s with a view to applying electricity to the traction of trains prepared the ground for its subsequent adoption. But it was the electrification of signalling that was to permit a real mutation in the organization of the system at an early stage. Thanks to electricity, transmission techniques experienced a spectacular break-through in the 1870s. Its application on the railways first concerned signalling equipment because growing traffic density on the lines was seriously affected by the flagrant and dangerous inadequacy of the signalling system. It was in the 1860s and the 1870s that different types of electric semaphores appeared, including the Siemens, which was later improved upon by French engineers. Based on the use of an electric magnet, it provided a means of moving what might be very considerable power from a distance by dispatching an instantaneous current. This innovation made the "block system" feasible and removed one of the major causes of the great railway disasters. At the same time the capacity of the networks was increased.

Four famous accidents occurred in 1876 (Chatillon), 1879 (Flers), 1880 (Clichy/Levallois) and 1881 (Charenton) causing 7, 10, 13 and 21 deaths. The signalling equipment was blamed for all four. A circular letter was accordingly sent out on 13 September 1880, and confirmed and underlined on 12 January 1882. It formally demanded the application of the block system beyond a certain level of traffic density and the "immediate and full harmonization of electric and sight signals". Such a program could only be achieved by the adoption of electric semaphores. Another circular letter ordered the gradual adoption of the Westinghouse brake, introduced on the Ouest network, for all passenger trains.

The electric semaphore and the continuous brake were complementary. The system gained in regularity and suddenly became far more coherent. These two innovations marked a decisive stage along the road to automation. It is interesting to note that American engineers tended to have far more confidence in automation than the French. At the International Railways Congress in 1900, in connection with the block system, the American representative had no hesitation in recommending the use of the automatic block which automatically put the signal to "halt" when a locomotive passed, whereas the French representative considered that it should serve only to "corroborate human action".[27] It would be beyond the scope of this chapter to describe the other forms of automation such as signal towers and the regulation of movements at stations. But it is important to note the appearance of growing concern over equipment management and over organization where the rotation of cars and running of trains were concerned. Rational methods of analysis were applied to traffic flows in the 1880s: "One might", concluded the author of an article on the distribution of cars in 1882, "find that the source of the great economies to be made in operating the railways lies elsewhere than in mechanical engineering".[28] Non-industrial engineering meant attempting to rationalize management by processing information.

All these considerations marked the first step in the efforts to emerge from the improvised empiricism of the Second Empire. But in this field operations were to come up against almost insurmontable barriers precisely because of the cumbersome and compartmentalized administrative organization described earlier. For instance it was not until the 1930s that the dispatching system was adopted on certain networks.

During the 1890s, when traction equipment for freight was underused, the engineers of the Nord and P.L.M. companies set about tackling the problems posed by the traction of passenger trains. The simultaneous increase in the loads and speed of these trains meant that locomotives were increasingly ill-adapted to needs. In 1885 research was undertaken on the Nord network with a view to fitting the double expansion (compounding) device already adopted for ship's engines to the locomotive. This research, carried out in coordination with the engineers of the Alsatian mechanical construction company succeeded, in 1891, in bringing out the locomotive 2121, the first of a generation of engines culminating, on the eve of war, in the Pacific 231. This breathtaking series of models owed a great deal to applied research, based on scientific principles, but also to the will to adapt them to operational needs. Compounding was subsequently also applied to freight trains, where the very strong pressure of traffic likewise entailed a very rapid increase in loading.

The technical system had thus reached a certain harmony while developing steadily. It was on the verge of a second energy revolution through electrification and would doubtless soon be endowed with rationalized management. Yet economically it was doomed because its directors could no longer keep costs under control. This inability was already perceptible before 1914. A balance had been achieved only by the systematic reduction of investments and could not last. This was one of the main causes of the serious social friction which culminated in the railway-workers strike of 1910.[29]

Problems of management and the march towards nationalization

The march towards the nationalization of 1937 began immediately after the First World War. There were three basic reasons why the directors of the networks gradually lost control of costs: state tutelage had deprived them of all means of independent action and was exercised to the benefit of other social actors; the organization had become less flexible; and the competition of other modes of transportation had developed whereas the companies' position was weakened by the highly labor intensive nature of railway technology.

A new convention signed in 1921 applied to the networks collectively. It settled the disputes provoked by the war and sought to create a new financial solidarity among the companies by setting up a "common fund". The aim was to arrive at a general equilibrium by using the surpluses of the prosperous companies, the Nord, Est and P.L.M. in order to finance the deficits of the others. The management of the networks remained autonomous. Joint bodies were set up but only enjoyed restricted powers. As it happened, only three years showed surplus balances over the period from 1921 to 1929. The crisis of the 1930s brought an unprecedented aggravation of the general deficit and after 1935 receipts failed even to cover operational costs.

The financial landslide caused by the crisis merely served to highlight the operational difficulties of the system. State tutelage had become too burdensome. The companies had gradually lost their freedom of action in a number of areas and particularly in that of setting rates. The State levied heavy charges on the companies without compensation, as such charges were the counterpart of a monopoly situation that had disappeared. The directors were no longer able to control the social system they had created. The former hierarchized and authoritarian pattern was slowly transformed into a guaranteed status system. The railway-workers' statutes after the war had organized the profession in a rigid fashion both as regards working conditions and career structures. When traffic collapsed in the 1930s the companies had no means of adjusting the numbers of their staff to cope with the new situation. What was more, the previous organization into skilled and functional technical departments was rusty. The companies had nevertheless made a great effort in the 1920s to introduce rationalization methods in their workshops and services and were alive to the inconvenience of their creaking bureaucracy. In 1928, the Nord network set up a commission on organization in an attempt to remedy the situation. In 1937, the commission noted: "It is difficult in an organization with independent divisions like in our network to avoid 'closed compartments' and to solve 'liaison problems'".[30] There are countless examples to illustrate the drawbacks of the lack of horizontal relationships between the sections: among others it led to tremendous wastage in the distribution of cars and orders for trains. Practically speaking, the entire management of the networks rested on a process of instructions and controls which functioned vertically. Instructions came down the hierarchical ladder, statements indicating irregularities in the departments went up. "These papers", a division inspector in Douai reported in 1931, "often ascend and descend the rungs of the hierarchy several times in succession. They are recorded and sifted on each floor, they take up a great deal of time and fill pages of correspondence before reaching the person who will supply the justification; then they engender more correspondence before arriving on the desk of the one who will finally assess the value of the justification given".[31] As this officer remarked, such conditions were very likely to cause irresponsible behavior.

The system of railway management therefore had two distinguishing features: a lack of proper communication between the departments and an incoherent codification of behavior. This erected a barrier and cut off the administrative divisions from everyday working realities. Regulations could not provide for everything and their very proliferation made them inapplicable, whereas the impermeability of the partitions prevented the dialogue required between those in charge. All this coincided to raise the costs of the railways at a time when they were open to increasingly aggressive competition. This explains the worried question asked by Javary, the Director of Operations of the Nord network, in 1927: "Can we still have a railway industry when labor takes 59% and capital charges 23%?"[32]. In 1932 Javary again remarked: "If the railway network did not exist already, it would certainly be only a third or a quarter of its present size".[33] He proposed reducing the number of stations on the Nord network from 755 to 68. According to his plan, automobile transportation should be used beyond those "gares centres" (central stations). But he never predicted that this transport mode would eventually compete with the railway over long distances!

The nationalization of 1937 was chiefly designed to permit economies in operations through the unification of the networks. Its justification resided in the idea, which had already appeared in the 1921 convention, that the particularism of the networks was largely responsible for their deficits. This idea was not wholly wrong, but it had the drawback of concealing other more profound causes. The system set up between 1852 and 1882 had functioned to the benefit of the nation and had been developed on the basis of steady innovation. There had been two technical systems indeed: that of the 1850s and that of the 1880s but both had worked efficiently. The difficulties were rooted in the system of organization, not in the technological one.

3 Conclusion

This historical analysis of the evolution of the French railway system allows the following conclusions to be drawn:

1. The emergence of the system was the outcome, in an atmosphere of technical and financial uncertainty, of a compromise between a government department anxious above all to maintain its control over paths of communication and bankers equally anxious to control the issue of securities and to govern an enterprise they hoped would be profitable. The solution adopted led to a mixed economy regime, combining the monopoly and the interest guarantee, in order to permit the financing of a network which has never been completed.

2. The first technical system, evolved in the 1850s, lacked coherence. It was gradually remodelled due to the pressure of traffic, which chiefly concerned the carriage of freight up to 1870 and passenger transport between 1880 and 1900. Traffic pressure also created a "demand for inventions" and brought the use of new technologies like steel and

electricity. This required a harmonization of the different components of the system. The dynamics of the system depended primarily on economic factors as the demand for innovation had been stimulated by the outside pressure of traffic. It should be remembered, however, that traffic density was due to the transport prices reductions and travel facilities provided under the system.

3. The technical system eventually ceased to generate high profits for its operators despite a significant rise in global factor productivity after 1886. The directors of the networks slowly allowed the control of costs to slip their grasp owing to higher demands from the State and poor adaptation in organization to meet the new needs of traffic management. The administrative structure remained unaltered and, though more and more cumbersome, failed to keep pace with changing realities. As a result a new demand for invention was engendered within the framework of railway operations with respect to the processing of information and total automation.

Notes

1. This paper has been translated by Barbara Thompson.
2. R.de Kaufmann, La politique française en matière de chemin de fer. Paris: C. Béranger, 1900.
3. A. Picard, Les chemins de fer français, Vol. 1. Paris: J. Rothschild, 1884, p. 18.
4. Y. Leclerq, Le réseau impossible. 1820-1852. Geneva: Librairie Droz, 1987, p. 136.
5. Picard 1884, op. cit., p. 101.
6. Cf. G.R. Hawke and M.C. Reed, "Railway capital in the United Kingdom in the nineteenth century". Economic History Review 22(1969).
7. Cf. Picard 1884, op. cit., p. 102.
8. Quoted in B. Zellemeyer, Les chemins de fer dans le bassin houiller de la Loire, 1820-1850. Actes du 98ème Congrès des Sociétés Savantes à Saint Etienne, 1973. Paris, 1975.
9. Cf. Picard 1884, op. cit., p. 106.
10. Cf. Picard 1884, op. cit., p. 127.
11. On the Paris - Saint Germain railway cf. B. Radcliffe, "The origins of the Paris - Saint Germain railway", Journal of Transport History 1(1972)4, p. 197-218. See also B. Radcliffe, "The building of the Paris - Saint Germain railway", The Journal of Transport History 2(1973)1, p. 20-40.
12. Picard 1884, op. cit., p. 252.
13. Ibid.
14. Picard 1884, op. cit., p. 253.

15 Annuaire Bloch 1847.
16 Cf. L. Girard, La politique des travaux publics sous le Second Empire. Paris: A. Colin, 1952.
17 F. Crouzet, "Essor, déclin et renaissance de l'industrie française de la locomotive 1838-1914", Revue d'Histoire Economique et Sociale 55(1977)1-2, p. 112-210.
18 On the Meudon accident cf. M. Baroli, Le train dans la littérature française. Paris, 1963, p. 63.
19 M.A. Curtel, Mémoire sur la fabrication et le prix de revient des rails. Paris, 1857. Quoted in F. Caron, French railroad investment 1850-1914 in R. Cameron (ed.), Essays in French Economic History. Homewood/Ill.: Richard D. Irwin, 1970.
20 Ch. Laboulaye, "Chemin de fer" in Dictionnaire des Arts et Manufactures, 5ème édition. Paris: Librairie du Dictionaire des Arts et Manufactures, 1881, p. 56.
21 A. Picard, Traité des chemins de fer, Vol. 3. Paris: J. Rothschild Editeur, 1887, p. 18.
22 Cf. F. Caron, Essai d'analyse historique d'une psychologie du travail: les mécaniciens et chauffeurs de la Compagnie du Chemin de fer du Nord de 1850 à 1910. Le Mouvement Social (1965)50.
23 Cf F. Caron, Histoire de l'exploitation d'un grand réseau. Paris: Mouton, 1973, p. 279.
24 Charles des Freycinet (1828-1923) was an Ingénieur des Ponts et Chaussées and a very important political man.
25 Caron 1973, op. cit., p. 125.
26 Caron 1973, op. cit., p. 392.
27 Congrès International des Chemins de fer. Sixième Session. Paris: 1900. Question XXV. Block System Automatique. Exposé no. 2 par M. Cossmann, p. 25.
28 Revue Générale des chemins de fer. Note sur l'utilisation des wagons au PLM. Mai 1882, p. 307.
29 F. Caron, "La grève des cheminots de 1910: une tentative d'approche" in Conjuncture économique, Structures sociales. Hommage à Ernest Labrousse. Paris: Mouton, 1974, p. 201-219.
30 Cf. F. Caron 1973, op. cit. p. 506.
31 Ibid. cf. p. 506.
32 Cf. ibid., p. 503.
33 Cf. ibid., p. 538.

CHAPTER 4
THE DEVELOPMENT OF THE GERMAN RAILROAD SYSTEM

G. Wolfgang Heinze and Heinrich H. Kill

1 Patterns of growth: An overview

Looking at the historical development of the German railroad network, one can identify four stages of development[1]. The primary stage (from about 1815 to 1840) involves the period when the original concept of building railways evolved and the first linkages of local importance were realized. During the following period (1841 to around 1875), connections between all the cities were built. During the third stage, which ended with the First World War, the existing lines were extended into rural areas and a feeder network was established.[2] The final stage, which continues into the present, is characterized by the decline of the German railroad system. This stage began with a period where railroad construction stagnated during the 1920s and led to the first closures of unproductive lines in the 1930s. Although this trend was arrested during the Second World War and its aftermath, these closures continued and were intensified during the last three decades. The connections remaining were adapted to advanced railway technology to meet present and future demands (Figure 1).

An analysis of other large technical systems indicates that this pattern of development is in no way unusual. A closer look at other transportation systems, as for example the development of inland water transport, inner city transit systems, or that of motorized road traffic similarly reveals four stages:

1. invention and isolated introduction (localized linkage),
2. demand-oriented construction (integration) - fulfilling only the needs of existing business centers,

Figure 1: Stages of German Railway Development

Period	General characteristics	German railways
Primary Stage (1815-1841)	Invention and isolated introduction (isolated linkages)	The original concept of building railways evolves (abroad): mostly private companies realize first linkages of local importance.
Secondary stage (1842-1875)	Demand-oriented construction (integration)	Connections are built between all major cities; governmental influence and even management grow.
Tertiary stage (1876-1913)	Demand and supply-oriented extension (intensification)	The state-owned but independent "Länderbahnen" in the German states are developed. Existing lines are extended into rural areas and a feeder network is established.
Final stage (1920 till today)	Maintenance oriented "cut back" (selection)	The Reichsbahn company is founded and the German railroad system declines. Railway construction stagnates in the 20s; first closures of unproductive lines follow in the 30s. More feeder lines are closed after WWII, the remaining connections are adapted to advanced railway technology.

3. supply-oriented extension (intensification) - supplying even the remote parts of the country motivated by the belief in equal access rights,
4. maintenance-oriented "cut back" (selection) - accepting efficiency as a basic principle and taking into account that different systems might complement each other. (In the case of road and freeway traffic the last stage has not yet been fully developed.)

One might therefore conclude that this is a general pattern in the development of successful technical systems serving a specific function. Whether system development will be successful is decided in the early stages of the process. The first decades of the German railroad system are also an excellent example of the decisive importance which the environment has for the system's chances of being successful. This paper will therefore concentrate on the first two stages of railway development we have identified in the beginning. We shall show in more detail that the development of the German railways was a sober economic affair of local interests. To overcome transportation bottlenecks, an existing technology was used in a new field and combined with another innovation of the era. The solution was found by engineers and travellers who were exposed to new ideas of solving problems tried by enterprises in England. The commitment of a few men led to independent innovative actions of a few communities. The overwhelming response of the social environment shows that the specific historical situation evoked transportation innovations and made the superior variant "railways" the superinnovation of bridging space. The polity only reacted to these events: Because it was a success, railway building became a favorite tool of governments.

The key elements of the German railroad era can be summed up in seven theses which will structure our argument.

1. Railways initially confronted rulers with a dilemma: On the one hand the railways were a very efficient or even necessary way to improve the economic situation and raise revenues, but on the other hand they increased the wealth and power of the bourgeoisie. Political response to the technical innovation was therefore ambivalent.
2. Time was ripe for a change because the old structures had reached their limits. In comparison to England, Germany was economically underdeveloped, though industrialization had already started. Population growth was high and labor migrated from the country into

the towns. Politically Germany was a loose confederation of 33 independent states (Deutscher Bund), which induced a deep desire for unification. Power was divided: Kings and the aristocracy held political power, while the bourgeoisie had the economic power. This historical situation with its various strains affected railroad development deeply.
3. In the initial phase there was a market for the railway, but most of the actors did not know it. The railway came on the stage when the general demand for transport had already grown enormously, but for most people improvements of existing transport facilities (waterways, roads, vehicles and traffic organization) seemed to be more than sufficient to meet new demands.
4. There were few who understood the new technical system and its rules. Most actors were caught in a cage of traditional thinking.
5. The fact that railway technology was introduced by transfer from abroad shaped the emerging system.
6. The state strongly influenced railway building, though private enterprises constructed and owned most of them.
7. The success of this technological innovation was its mass effect: Rising utilization led to profits and cost reductions, which triggered a positive feedback making the railways the leading sector of German industrialization.

The railway was more than a new means of transportation with higher capacity. It opened new psychological, social, economic, political, and military dimensions, maybe comparable to the first flight across the Atlantic or the first landing on the moon. Until the advent of the railway, transportation was mostly dependent upon horses, the force of wind and the speed of running water. Travelling was a tough business, costly, slow and risky. Horses and carriages were something for the rich and powerful. In Germany the ordinary man went over long distances on foot. The railway changed all that. This new way of space-bridging and mobility led to a new perception of space, distances, speed and time. And this new means of transportation was not only something for big cities but could be used by everybody to go nearly everywhere and in all directions. The world shrank and the multi-state system became an anachronism. At the same time, however, new kinds of accidents caused fear, and technology was felt as a threat.

2 The historical background

In the beginning of the 19th century, when the Napoleonic Wars were followed by political restoration, Germany re-established its multi-state system. Only a few of the reforms of the Liberation Wars remained in force; the Stein-Hardenberg Reforms, in particular, were hardly affected. But the bourgeoisie did not receive the promised share in political responsibility. Only in some minor German states were parliaments set up so that they could be called constitutional monarchies. This was still the political situation in the first phase of railway development. The few political changes that did take place, while containing the seed for substantial future changes, did not actually affect the political balance of power at the time. But they affected the chances of the railways. Especially the foundation of the "German Zollverein" in 1834 must be mentioned here. Though this simply meant that many of the German states (with the important exclusion of Austria) adopted the Prussian trade and customs regulations[3], a market big enough for a substantial growth in trade and commerce was created. Apart from that, the nationalistic liberal movement that ended in the suppression of the 1848 revolution and the enfranchisement of all three classes was important, as it helped to increase especially passenger traffic within Germany.

When the idea of building railways arose, the governments found themselves in a dilemma with respect to this new means of transportation. On the one hand, it was clearly seen by most officials that railways were a very efficient or even necessary way to improve the economic situation and the competitiveness of the country - and hence it was a way to raise states' revenues. On the other hand, the spreading of the railway network had two important disadvantages: first it contributed to the wealth and power of the bourgeoisie who built it because the absolutistic governments did not have the money and were unwilling to borrow it because of the political obligations attached. Secondly, the railways improved the mobility of the people and therefore the diffusion of new ideas beyond a point still regarded as tolerable. This is why early on the Prussian King Friedrich Wilhelm III did not even permit the privately financed building of a railway. Later most states and their monarchical governments publicly supported the idea of building railways, but in fact often worked against their actual construction[4].

110 G.W. Heinze, H.H. Kill

Figure 2:

THE DEVELOPMENT OF THE GERMAN RAILWAY
NETWORK 1842 – 1913

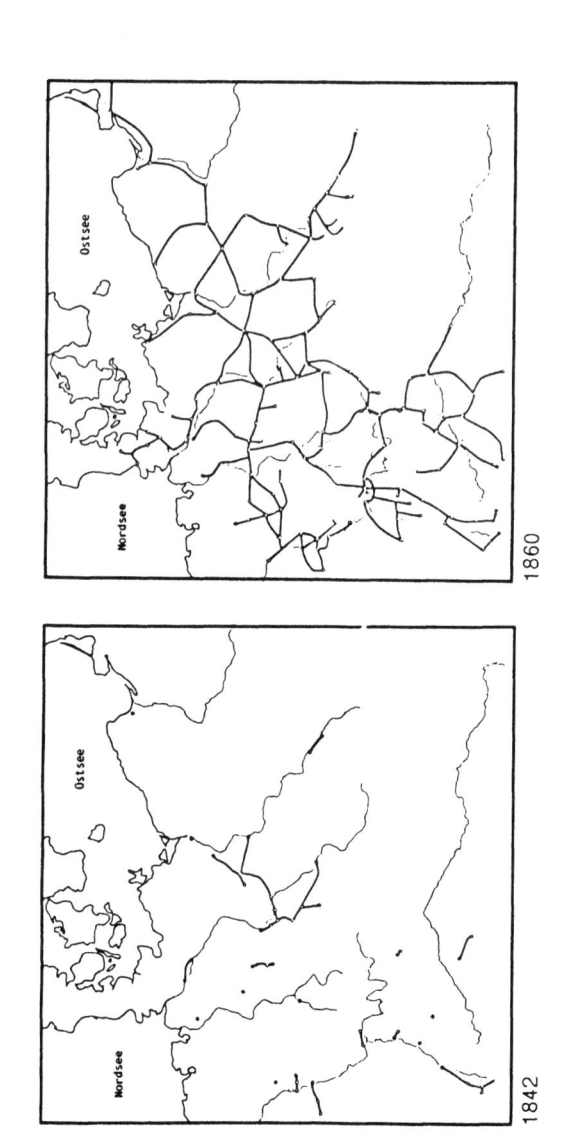

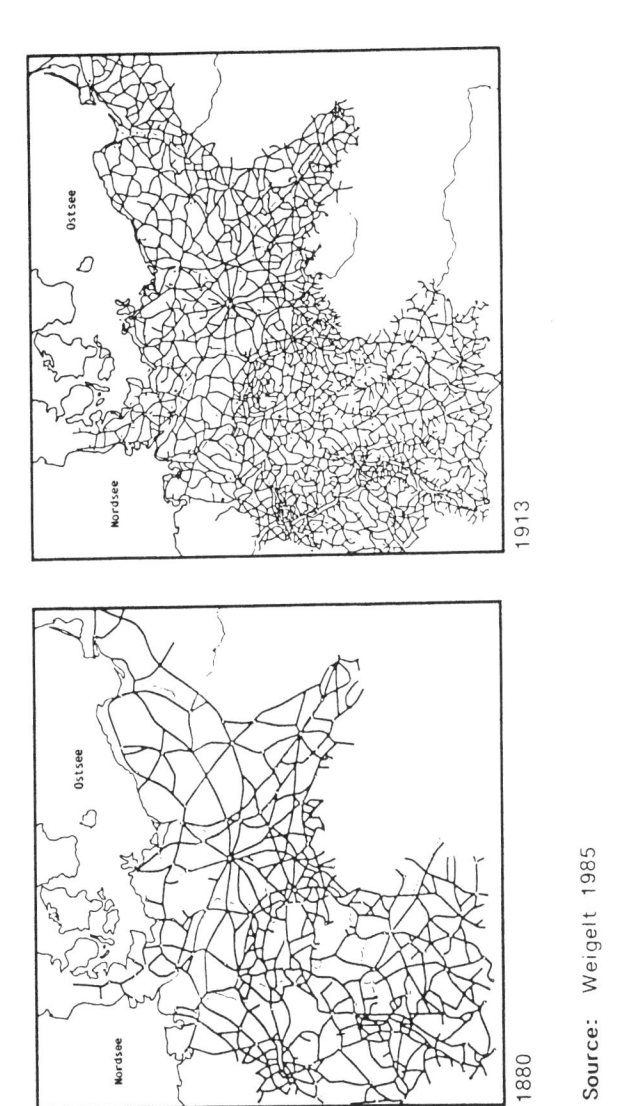

The German Railroad System

1913

1880

Source: Weigelt 1985

The fact that the railways, despite many obstacles, actually spread in Germany with a time lag of only 10 to 15 years as compared to England shows that other - mainly socio-economic and technical - factors were favorable to such a development.

While the political framework stayed quite static, the economy, transportation technology, and technical development generally were undergoing rapid changes, creating pressures for structural adaptation. Until the 1830s, the German economy was clearly dominated by agriculture, with up to 80% of the workforce engaged in food production. Only in Saxony and in the Prussian Rhineland was a majority already engaged in trade and manufacturing. But even there, manufacturing took place in small plants with few workers and modern machinery was based only on a very small scale usage. From today's point of view the use of steam power is regarded as the symbol of industrialization. But the few German steam engines - not even one tenth of the number they had in England - were mostly used for water pumping, especially in the coal and salt mines but also in factories which had water powered machinery.

In England the use of machinery and steam engines, together with modern forms of management and production organization, led to an immense increase in productivity and consequently to a decrease in the prices of products. The German craftsmen therefore could not compete with the British factories, even more so as the influx of British goods was not really hindered by taxation during the first years after the end of the "Blockade of Britain". The introduction of large-scale production in Germany after the 1830s was mainly based on two factors: the introduction of foreign methods of mechanized manufacturing and the introduction of new ways of financing which led to the founding of joint-stock companies. In order to raise the required capital these companies offered shares to the public. This method was originally used by the trade companies engaged in trade with the West and East Indies. But in these companies the share-holder also had to account for the losses of the company. The first modern joint-stock company enabled the shareholders to share the companies' profit, but limited the risk to the value of their share. In Germany such a company was founded for the first time to finance a coal railway, where the wagons carrying coal were drawn by horses.

Technological development during this time was generally characterized by the shift from medieval technologies based on water-powered wooden machinery to modern ways of production using steam-engines and advanced machinery. Along with improvements in the iron, steel and mining industries, German mechanical engineering developed. The most important factors in this development were the rising level of general education, and the installation of model factories with advanced foreign technologies. The educational reform policy led, among other things, to the foundation of the first German technical universities in Prague in 1806 and in Karlsruhe in 1825. In Prussia the "Königliche Gewerbeinstitut" was founded in 1821 in Charlottenburg near Berlin (this later became the "Technische Hochschule"). This kind of modernization was supported by the government[5]. Prussia, the most important German state, was large enough to need a bureaucratic administration and it employed many modern, economically thinking professionals.

3 Rising transportation demand and the difficult search for a niche

The now growing industrial production with its even faster growing demand for transportation revealed the weakness of the existing transportation system. Besides the trading companies, which were traditionally interested in good transportation facilities, the management of the heavy industries was now interested in improving the transportation sector.

During the Middle Ages it was above all improvements in vehicles that maintained or even enabled the increase in the volume of transportation.[6] Paved roads in general did not exist. Only at specific locations, e.g. river crossings, swamps or mountain slopes, was construction work regarded necessary. Together with the development of national economies in Western Europe, the 18th century saw the first nationwide road building since the Roman Empire in Europe. The importance of good traffic connections as a prerequisite for the development of a nation was realized especially in multi-state Germany. In 1779 Christian von Lüder demanded a network of high quality roads connecting all the important cities of central Europe. Interestingly, this plan asked for "Chaussee" connections very similar to the later network planned for railroads by Friedrich List[7] to promote a unified Germany. They are also almost identical with the 20th century plans for the Reichsautobahnen. None of the three networks was actually built on the basis of these plans. Only independent connections were built, due to the influence of state and local particularism, the importance of profitability as a criterion and, in the case of the Autobahnen, because of strategic planning. Nevertheless, over time the networks evolved to a state surprisingly similar to the one originally conceived.

Despite the fact that new roads and canals were built during the time of Mercantilism and especially during the Napoleonic occupation, traffic connections as a whole were not in a good shape. This became obvious when the demand for transportation facilities increased in the 20s and 30s of the last century. Nobody doubted the need for an improvement of the transportation system, but as nobody could predict that the increasing demand for transport would continue, the need for a completely new transportation mode was not evident. For most people improvements of the waterways, the roads, the vehicles and of the traffic organization seemed to be more than sufficient to meet future demand. This clearly worked against speedy railway development. As late as 1834, a canal was chosen to connect the Danube and Main. Besides the advantage of the canal for transporting timber - Bavaria's most important export product of that time - the people responsible for this decision had two main arguments against the railroad. First mistrust in the reliability of the new machinery, and secondly disbelief in a steadily growing demand for transportation.[8]

Furthermore the power of steam engines could be used more easily on waterways than on land. In those days steam engines were quite big and heavy and needed huge amounts of coal, and it was much easier to put this additional weight on a boat than on a wagon. Therefore, as early as 1816 a steamship was used for a shuttle service by the Prussian Post Office between Berlin and Potsdam. Steamboats started to operate on the Rhine in 1822. They became so successful that in 1830 the "Rheinisch-Preußische Dampfschiffahrtsgesellschaft" alone had 27 steamships in service between Köln and Mainz.[9]

Nevertheless, the larger part of the investments in traffic infrastructure went into highway building. The improvement of these new roads was so remarkable compared to the old ones that they were called "artificial streets". In the three years from 1805 to 1807, more than 5,000 km of highways (Chausseen) were built or improved in Bavaria, and in Prussia the length of the highways doubled between 1830 and 1848 to 15,000 km.

The improvements that followed from these measures (and especially from the improvements in traffic organization) were quite considerable. Cargo as well as passenger and mail transportation time was greatly reduced and capacity rose in the same manner.[10] It can therefore be said that there was no general or even national interest in a railway system when its building began. Beside the more visionary imaginations of some poets and scientists, there were only the local interests of tradesmen and manufacturers in different cities who wanted the railroad for improving communication and transit of their goods which could not be achieved by other means.

The first article on a (planned) German railway was published in 1825 and was written by the entrepreneur Friedrich Harkort. Although he first described the advantages of a railway connection between the North Sea and the Rhine (to avoid the Dutch Rhine customs), his true intentions lay elsewhere[11]. In Wetter by the Ruhr he had founded a factory (the later DEMAG) with imported English machinery and English workers. The connections to the nearby coal districts were very poor. Coal was even transported on horseback. Therefore he wanted to improve the transport of coal from the mines to the "Bergisch-Märkische" industrial district. Because of his initiative a test railway using the Palmer principle and horsepower was built in Elberfeld in 1826. This one-rail track was followed by several - now two-tracked - railways in the Ruhr Region that connected coal mines with neighboring cities.[12] Their length was usually only a few kilometers, but on these railways one horse could pull more than a tenfold of an ordinary carriage. Similar to this case, also in the other regions it was individuals who pushed the idea of railways forward[13], and quite often projects became delayed (or even suspended) because of the death of one man.

At the beginning of the 1830s, railway projects had sprung up in nearly all major German cities.[14] Generally these projects were advanced by citizens who intended to improve their city's position in the economic competition with other cities. The promotion of the railway as a system was not the intention of these system-builders.

Beside these entrepreneurs who were only interested in one particular railway line, Friedrich List (1789-1846) was obviously the only man with a great plan and with an understanding of the whole system[15]. He failed because he did not adjust to this disjointed incrementalism. A design for Germany as a whole was also doomed because general interest was lacking. For many reasons List was an outsider. As a self-made man with liberal ideas, not rich enough but with a hot temper, he did not belong to any social group of political influence. The nobility called him a "revolutionary", the merchants "office-hunter" (Postenjäger) and the academic establishment "agitator". His personal tragedy was that he functioned as a catalyst without reaping any personal benefits. His efforts to see Germany as a whole are comparable to our attempts to think European. Fixed on his aim, he even opposed "small steps" that were improvements of the status quo; he did not understand that complex systems have to prefer evolution to revolution because big steps lead to extreme rates of change in other subsystems and thereby endanger the whole system. Not supporting incremental changes, List in fact played into the hands of the reactionaries. The result was not a grand solution, but a lagging disjointed incrementalism of German railway development. The great names of German railway history are all names of losers. Looking backward, von Baader had invented the wrong system, Harkort went bankrupt and List committed suicide[16]. Later, however, the decentralized network structure proved to be an advantage. According to the unanimous judgement of regional scientists, German particularism led to less regional disparities and higher structural adaptability, lowered the costs of regional policy measures and improved social consensus.

4 Early railway construction: The effects of technology transfer, capital needs, and state regulation

The early technological development in the transportation sector abroad gave Germany the advantage of being able to adopt complete, proven and functioning systems which had already passed their teething troubles. The most influential model was the technologically and financially successful railway between Manchester and Liverpool which opened in 1830. One highly significant consequence of this mode of innovation by transfer was the introduction of the standardized gauge[17]. Only the state railway of Baden used a larger gauge (which was probably better suited for the flat Upper Rhine Valley) in the beginning, but changed to the standard in 1855. But technology transfer also had disadvantages. The English systems were considered to be in optimal shape and improvements were not thought possible[18]. The plans for the capacity and the routes of the early German railways therefore reflected the state of technology England had achieved several years before.

In this context, the famous "Ludwigs-Eisenbahn" between Nürnberg and Fürth was not the symbol of unique pioneer work in the new technology. Two years before its opening, the railway had only been one of many projects, but luckily it was finished very quickly. The society promoting this railway, the Nürnberg-Fürther-Eisenbahn-Gesellschaft, was founded in the fall of 1833, it got the concession on February 19, 1834, and it opened only one and a half years later (1835). The advantage of the Ludwigs-Eisenbahn was that all obstacles that usually delayed the realization of such projects could be overcome quickly. Both city councils were in favor of the project and most of the money necessary was provided by citizens. The cost for the complete railway, including locomotive and wagons, was only 122,000 talers (compared to the average of 150,000 to 600,000 talers just for the construction of the same length of infrastructure elsewhere).[19] The royal privilege that was given with the concession made it even easier to raise the money, especially after the King had bought two shares. The terrain did not cause any difficulties (aside from the question of expropriation); neither bridges nor tunnels were necessary, and all streets were crossed at the same level [20]. Finally there was no competitive organization fighting the project.

All other railway projects in Germany were hindered or even caused to fail by the problems connected with (1) finances, (2) route-finding, and (3) concessions[21]:

1) The financial problem had to be solved first. The amounts of money necessary for building a railway were extraordinarily high. The possible financiers had to be convinced of the success of their investment. In the beginning, this was very difficult because no railway had yet been built in Germany. Besides, profits could only be expected after several years. Therefore many of the potential investors were skeptical about the railway's technical performance and its financial profitability. But this problem was largely resolved when the "Ludwigs-Eisenbahn" was built and when it proved to be economically successful: After its first year, the company paid a dividend of 20%.

2) Routing was not so important when discussions about a new railway started, as most railways were planned as connections between two major cities. But when the actual construction began, many decisions had to be made. Which places between the two cities should be con-

nected to the railroad? In which part of the city should the terminal be? Where should rivers and mountains be crossed? How could important factories and army forts be reached? How could towns and noblemen who declined to participate be circumvented? These kinds of issues often delayed construction.

3) Though private enterprises constructed most early German railways, the state could refuse to grant necessary elements: the concession, the law to expropriate grounds, and the interest guarantee. The state had a seat and a vote in the administration of the companies. Soon the various state governments realized that they were indispensable to get a railway network which included many necessary, but non-profitable lines. In spite of this awareness, most German states accepted privately owned railways under strict state control. Requests for a concession where often promptly met, but generally these concessions were tied to a number of preconditions, so that construction work could not start at once. Besides this, the government sometimes promised concessions to more than one party or withdrew concessions because of minor violations of the stipulated conditions.

Many of the first railway lines proved not to be profitable. Railway building became much more expensive than expected. Only Paul Camille Denise, a German engineer who had also built the "Ludwigs-Eisenbahn", could say that all his railways were profitable. He built simple but solid tracks, which meant that the construction costs were relatively high at the beginning, but the long-term operating costs could be kept low. Higher construction costs are soon past and forgotten. Many of the other companies were forced to resort to the interest guarantees of the state.

5 After a difficult start, a quick take-off

Due to these difficulties all the other railway projects except for the Ludwigs-Eisenbahn remained in a planning or even discussion stage, until the positive results of the latter became public by 1836 and 1837. From then on, the interest of governments and of investors in railway construction was big enough to push the other projects forward[22]. Within 5 years, from 1837 to 1841, 680 km of railways were opened. The first state railway connected Braunschweig with the Harz mountains (30 km). All the other railways connected nearby cities: Berlin and Potsdam (25 km), Leipzig and Dresden (120 km), Frankfurt/Main and Wiesbaden-Mainz (40 km), Düsseldorf and Elberfeld (Wuppertal) (30 km), Munich and Augsburg (70 km), Berlin and Frankfurt/Oder (100 km), Magdeburg and Leipzig (110 km), Mannheim and Heidelberg (20 km), Cologne and Aachen (70 km), Hamburg and Bergedorf (20 km). Aside from these lines, some extensions were also in progress. These city connections became later the nuclei of the polycentric German railway network.

The railway boom was accompanied by several speculative projects which soon proved to be unrealistic (at that time) and many people lost their money. Often stock-companies were founded and shares sold before the railway got the concession. During this boom the costs of building the railways rose immensely. When the "Westfälische Landtag" allowed the foundation of the "Köln-Mindener-Eisenbahn Company" in January 1831, the 230 km connection of the rivers Rhine and Weser was estimated at 600,000 talers. In 1836, when the company was finally constituted, this calculation had risen to 4.4 million. In 1840 it had risen to 5.6 million. When this railway finally opened in 1847, construction costs amounted to 13 million talers, i.e. about one fourth of the annual budget of Prussia of that time.

The main reason for the delay in the construction of this particular railroad was its length. It was the first railway that was intended not only to connect two cities, but was planned as the connection between two distant rivers, the Rhine and Weser. This railway can thus be called the first German long-distance railway.[23] It traversed not only densely populated commercial areas, but also rural areas. For such areas it was very difficult to raise any money, because it was not certain whether the railways could find enough passengers and cargo there. During the time it took to realize the Rhine-Weser railway, the pioneer stage of the railroad systems development had ended.

By 1840 the importance of the railroad system as a whole for the nation or the individual states had become obvious. Only one year after the opening of the Köln-Mindener railway, it became the center part of the first transnational railroad from Frankfurt/Oder to Aachen (with connection to Belgium). Although Prussia adhered to the concept of privately owned railway companies for another 40 years, the companies now came under very strict state control. Except for Prussia and Saxony, all the other states in Germany now decided to build and operate railways themselves - without necessarily taking over or prohibiting all privately owned railways. At the beginning of the 1840s the German states made plans for their railroad networks. These plans were no longer dominated by the potential traffic between two points. Some states planned railways in order to influence the international trade routes - or in fact to take them away from the other German states' territory.[24] One of the main aspects of railway planning was now to extend the accessibility to all provinces to give them better chances for economic development. But despite these official plans, the main reference point - where to start building railway - was the existing traffic. Therefore it took only ten more years to finish a skeleton of railway tracks in Germany. In 1855 Germany had about 7,500 km of railroads. These lines followed traditional traffic patterns of the last thousand years and corresponded very much to the road network plan of Lüder, or to the railroad plan from Friedrich List.

Although the main purpose of building the railways had been the transportation of goods, in the 1830s and 1840s they primarily attracted passengers. The manner in which the railroad network was developed made it very difficult to attract cargo. Most railways opened operation before they got to the final destination, usually when the nearest town was reached. Beside the fact that long-time treaties between merchants and haulers quite often forced shippers to use road transport, there was also the problem of having to transship the cargo between different lines as well as between different transport modes. As long as the railway could not cover the whole transportation route, repeated reloading often took more time than the supposedly fast railway saved. In contrast to cargo, passenger figures soon exceeded their predictions by far. The predictions were based on the number of people then using the carriages between the cities where the railway was planned. Most of the additional passengers were former pedestrians, belonging to a social group that did not use public transport before, mostly because it was too costly. Riding carriages was hardly faster than walking. Suddenly by using the railway one could save so much time that one could get to the next town, have one's business done and return the same day. So even those who usually walked could save money: The price for a 30-km return ticket was 2.40 marks in third class and only 1.60 marks in fourth class, which is less than what one usually had to spend for lodging and eating, not considering the one day saved for working and the cost for shoes and clothing.[25]

The development of the regular passenger transport to such an unprecedented and unpredicted degree is one of the best examples of the difficulties of future oriented technology assessment[26].

In cargo transportation the importance originally conceded to general cargo was very soon displaced by bulk cargo. But in order to compete with the water-transport for this kind of cargo, two more developments in the railroad system had to happen[27]. On the one hand the price of cargo carriage had to drop almost to the level of water-transport, and on the other hand the capacity of the railways had to be increased in order to match the amounts of cargo the increasing industries demanded. These developments cannot be seen in a network map.

Most of the coal used in Germany at the time came from England and was carried by boats at a rate of about 1 taler-pfennig (i.e. one hundredth of a taler) per ton kilometer. As the railway companies did quite well with the existing passenger and cargo transportation, they saw no necessity to reduce their freight rates, which were about 10-15 taler-pfennigs per ton kilometer. Due to the involvement of the Prussian Secretary of Commerce, August von der Heydt, and because of a cooperation among the coal mines in Upper Silesia, the railway company serving this area in 1849 offered the first one-pfennig-tariff train to Berlin. Within a few years these cheap trains that carried only coal - which was an innovation as well - became one of the main source of revenues. The share of hard coal in the cargo transport volume of the Prussian railways rose from 1% in 1850 to 14% in 1860 and to 31% in 1875. The total coal transport in Prussia reached 1,012.8 million tkm in 1865 (45.1% of the total cargo transportation) with a freight of 37.2 million marks (29% of the railways' cargo revenues)[28]. In other words: although they did not recognize this in the beginning, it was the railways themselves that induced the low value mass transport demand they were best suited for.

Yet reasonable prices alone were not enough to initiate this development. The railroad network had to be modified to match this rising transportation volume. Even when all major destinations where reached by the railroad, cargo still had to be transshipped. The different lines terminated in the outskirts of the cities. Each line had its own station and even if the stations lay sometimes quite closely together, the lines did not connect. For the passengers, transfer to the next station might have been annoying, but for the cargo this was a real obstacle. Furthermore, most of the first railways were single tracks, and the stations along the line were often simply places where the trains stopped, which meant that when stopping at a particular station the complete line was blocked for the time necessary to load and unload cargo.

Long before Germany was covered by a complete railroad network, the improvement of the core network was carried out. In Breslau the stations of the Upper-Silesian railway and the Lower Silesian-Märkische railway were connected in 1850 (this was one of the preconditions for the previously mentioned coal train to Berlin). In the following year the Prussian state built the semicircular Connection Railway in Berlin. Similar connections were built in Leipzig in the same year, and in Dresden in 1852. In Cologne, only in 1859 was the Rhine crossed by a bridge, connecting the Rhine-Weser railway terminal in Deutz with the Köln-Aachen-Antwerpen terminal on the left side of the Rhine.

Meanwhile the established lines were improved to allow for higher speed trains, second and third tracks were constructed, and switches were installed in and between the stations. This allowed the bypassing of faster trains[29] and facilitated direct railroad connections to factories, or even to agricultural and mining facilities.

6 The end of the introduction phase

By the early 1850s it had become obvious that the railways were utilized to the greatest effect when they enabled cargo transportation without the need to transship. As the railways were so superior to the traditional ways of transportation, long loop-ways (detours) were accepted in order to keep the cargo (or the passengers) on the rail. Although the railroad network of that time covered most of the traditional trade-routes, some connections were still missing. In the first 20 years of railway history several projects had failed because of a specific German feature: Political particularism had produced enclaves and some absurd borderlines, and traffic routes between two cities in one state would therefore have to cross borderlines, but permission to do so was usually denied. That made such railways financially unattractive. But soon the governments saw the importance of interstate and international traffic for the states' own industry and commerce and ensured the future of the railways by treaties. Before, it had taken several

The German Railroad System 123

Figure 3:

MAIN INDICATORS OF THE
GERMAN RAILWAY SYSTEM'S GROWTH

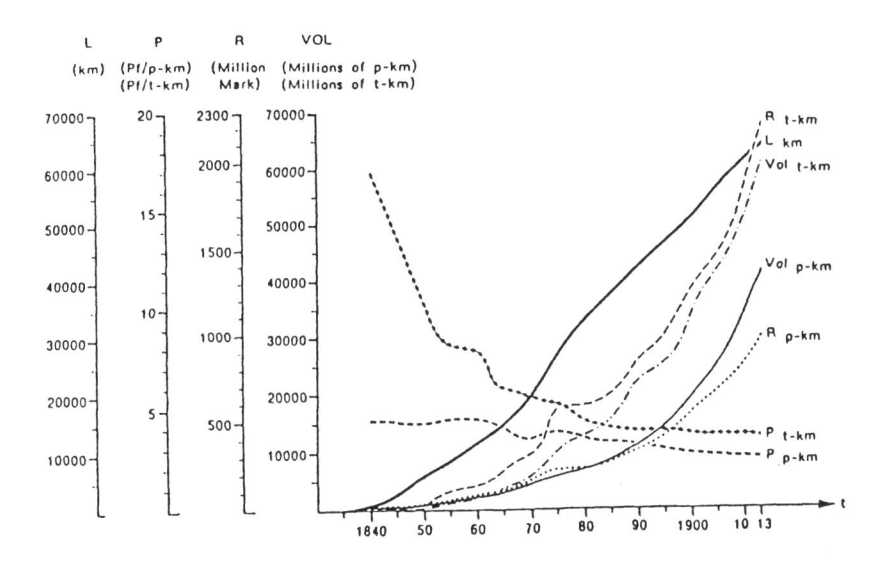

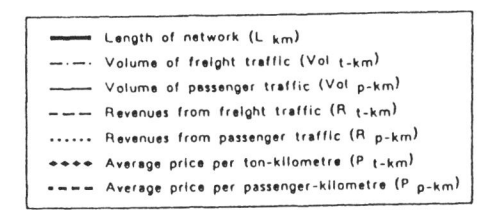

L	P	R	VOL
(km)	(Pf/p-km)	(Million	(Millions of p-km)
	(Pf/t-km)	Mark)	(Millions of t-km)

Curves labeled: R t-km, L km, Vol t-km, Vol p-km, R p-km, P t-km, P p-km (time axis 1840, 50, 60, 70, 80, 90, 1900, 10, 13).

Legend:
- ▬▬▬ Length of network (L km)
- —·—·— Volume of freight traffic (Vol t-km)
- ——— Volume of passenger traffic (Vol p-km)
- — — — Revenues from freight traffic (R t-km)
- ······ Revenues from passenger traffic (R p-km)
- ♦♦♦♦ Average price per ton-kilometre (P t-km)
- ▪ ▪ ▪ ▪ Average price per passenger-kilometre (P p-km)

Source: dtv-Lexikon, Bd. 5, München 1974; Fremdling, R.: Eisenbahnen und deutsches Wirtschaftswachstum 1840-1879, Dortmund 1975; Huber, P.: Die deutsche Eisenbahnentwicklung: Wegweiser für eine zukünftige Fernschnellbahn?, DFVLR-FB 78-25, Köln 1978; Handbuch der deutschen Eisenbahnstrecken, Mainz 1984; Voigt 1965

years to reach an agreement between two German states regarding one short railroad. Now within a few years, general treaties between the German states and their neighbors were signed covering traffic connections between their capitals and between important trade centers.

This governmental willingness to improve the railroad network soon resulted in the closing of still remaining missing links. In Prussia the number of concessions granted increased, and in Bavaria the state even allowed private companies to build and operate railways. One important reason for this increase in concessions was probably the intention to intensify competition between the railways; in Germany the state railways played the same role as competitors as the private companies did in Britain or in America. By 1865 all the old trade routes had rail track and each city and mining district could be reached without long detours. The length of the railroad network had reached nearly 14,000 km. In the early 1870s more than 1,000 km of railroads were built annually, and between 1865 and 1875, 13,700 km. That is about the length that had been built in the 30 years before.

With this completion of the mainline-network, the integrating stage of the system's evolution had nearly ended. The connections finished during the 1870s had all been planned in this period. It is interesting to note that this integration of the railroad network coincided with the unification of Germany. After the "Reichsgründung" of 1871, two factors led to the further expansion and subsequently to the intensification of the network. For one thing, earlier railway lines had been built from town to town so as to collect as much traffic as possible. This led to remarkably loopy ways for the traffic between the terminal cities. Now, competitive companies established direct linkages between major destinations. Secondly and more importantly, competition resulted in the building of financially doubtful railways into rural areas. However, such lines were built not only out of mere speculation, but also because of the state's intention to grant improved accessibility to every region. So, although the nearly 30,000 km of 1875 covered Germany with a complete railway network, railway building continued at the rate of about 1,000 km annually (Figure 4).

Quite often railway building was requested by the towns not yet connected. In the wake of industrialization, German towns and communities situated near the railway network which were not yet serviced made strenuous efforts to establish linkages, hoping to help local producers. However, the opposite often occurred. Extension of the

The German Railroad System

Figure 4:

THE PROVISION OF RURAL TRANSPORT BY LOCAL RAILWAY LINES AND BUS LINES IN BAVARIA 1840 - 1980

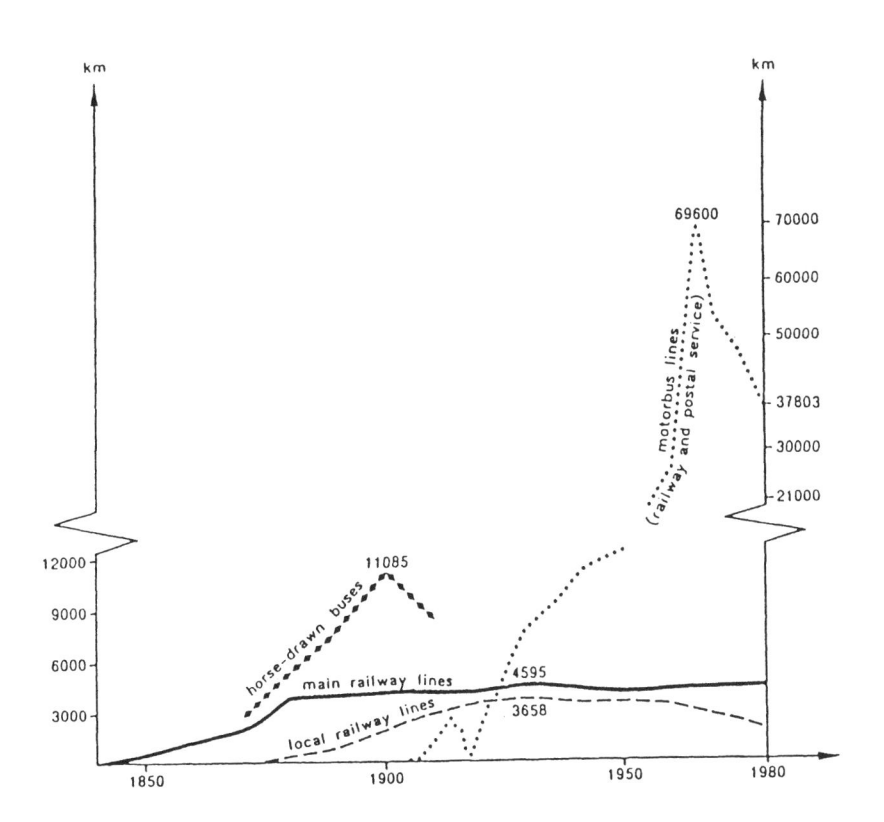

Source: Weigelt, H.: "Der öffentliche Linienverkehr in der Fläche - aus der Sicht der Deutschen Bundesbahn"; in: Die Bedienung ländlicher Räume - eine Aufgabe aller Verkehrsträger; Schriftenreihe der DVWG, Band B 64, Köln 1983, S. 84

railway network of the State of Bavaria to these zones caused an economic decline, whereas the zones of industrial growth continued to expand. Nearly all those production sectors and enterprises in the declining zones which had been mentioned in the petitions as justification for the requested rail linkage were forced out of the market in the long run. Economic development was marked at the junction points of the main lines and at the terminal stations in the towns, where transfer to road haulage was effected[30].

7 Nationalization of the railways

During the intensification stage that followed the completion of the mainline network after 1875, organizational changes also occurred. This period is characterized by the integration of the public railways, which were juridically and administratively independent, and the private railways to form the "Länderbahnen", i.e. one state railway for each of the bigger German states.

Although nationalization of the railways had been demanded since the beginning, a necessity to do so was never seen by parliaments and governments. Especially the already mentioned pioneers of railway building, Harkort and List, pleaded for state railways in their first papers. After the separate railway companies had evolved in Germany, they gained most of the advantages of a unified railway by voluntary collaboration. Despite their competition, the railway companies had started to cooperate quite early. Already in 1846, ten of the 17 Prussian railway companies of that time founded the "Verband der Preußischen Eisenbahnen". The aim of this association was to standardize the technical equipment, the rolling stock and the overall dimensions of bridges and tunnels so that trains and especially wagons could use the tracks of different railways. Furthermore first regulations for standardized tariffs were discussed. In 1847 all railway organizations in Germany founded the "Verein Deutscher Eisenbahnverwaltungen" in order to extend these standardization efforts, and before it was possible to travel from one side of Germany to the other in the same train, one could make such a journey with several trains but one ticket.

While this system worked very well in peacetime, the disadvantages of such a multicompany railway system became evident during the

French-German war of 1870-1871. Coordination problems led to delays in troop and material transports. These experiences and the spirit of the newly created German Reich renewed intentions to unify the railways as well. But similar to the case of the German trade policy at the beginning of the century, a coalition of economic liberalism and political particularism obstructed a solution for Germany as a whole. To prevent the possibility of a Prussian dominated German state railway, the medium sized states of Bavaria (1875) and Saxony (1876) nationalized all important railways in their territory. Prussia, and here first of all Bismarck, therefore concentrated unification plans on Prussia's own territory and on small neighboring states. Until 1887, Prussia bought all private railway companies that operated main lines in its territory.

This polycentric concentration process excluded only small private companies that served secondary lines of local importance. At the turn of the 20th century, 59,082 km of the 63,794 km of German railways belonged to 8 state railways. These states were Prussia, Hesse, Saxony, Bavaria, Württemberg, Baden and the small but opinionated duchies of Oldenburg and Mecklenburg-Schwerin.

When the railway had reached this heyday of its development, it was not just a new means of transportation with a higher capacity. It had become a new dimension of space- and time-bridging[31]. The railways became the necessary precondition of economic development and proved to be very profitable themselves. The annual revenues of the 8 state railways averaged more than 1,000 million Reichsmarks at that time.

Because of its importance, the railway was strongly influenced by world politics. It was a direct result of defeat in the First World War that a national railway company was finally founded in 1920. Forced by the peace conditions of the allies and the new constitution, the "Länderbahnen" were unified to the "Deutsche Reichsbahn". As the new German Republic could not afford to take credits that were necessary to rebuild the damaged railways, the state-owned "Deutsche Reichsbahn" was converted into a legally, administratively and financially independent company, the "Deutsche Reichsbahngesellschaft".[32]

Besides paying off its own debts, this company was forced to produce 600 million gold marks (i.e. about 150 million gold dollars) annually to redeem German reparations to the Allies. In the years from 1925 to 1932 the Deutsche Reichsbahn paid 4.18 billion gold marks, although in 1928 for instance it still had to make repair investments that added up to 2.5 billion Reichsmarks. So, at the beginning of the final stage of its development, the railway not only faced the competition of a new transport technology (i.e. the automobile), but also severe financial problems.

8 Summary and conclusion

Much as in the introduction of other new transportation systems, the railway had to face the strongest opposition in its initial phase. The groups that opposed the railway stood to lose what they had achieved with the old and well-known technology. The first supporters of the railways therefore pointed out the advantages of the railways for the extension of agricultural production and for accessibility to the spas (like Baden-Baden). As the Postmaster General opposed most of the early railways, mail had to be carried without payment and the railways had to compensate for the losses of the mail coaches. Because of these obstacles and because of the dominating influence of the traditional technologies (especially inland navigation), the railways found their first employment in niches or when really no other means of transportation was appropriate. As the introduction of the railway took place during a period of general growth, the limits of the old system's capacity became evident. Other favorable conditions were the maturity of the new technology at the right time and the personal engagement of open-minded entrepreneurs - men who were familiar with the new technology, convinced of its success, and who had the economic and political knowledge to push its introduction (Figure 5).

After the railway had been introduced and had shown its operational abilities, other cities and states reproduced it in manifold ways. Because of this, the railway could extend and at the same time modify the system, i.e. the growth of the new technology actually enabled the structural changes that had become necessary. When the railway had become the dominant transportation system in the 1870s, its field of operation spread over the whole transportation sector. Because of its superiority compared to other transportation modes and its economic success, railways were built to almost every town and were used for nearly every transport purpose. This dynamic mechanism of success

The German Railroad System

Figure 5: THE GERMAN CASE: A SIMPLIFIED MODEL OF RAILWAY DEVELOPMENT

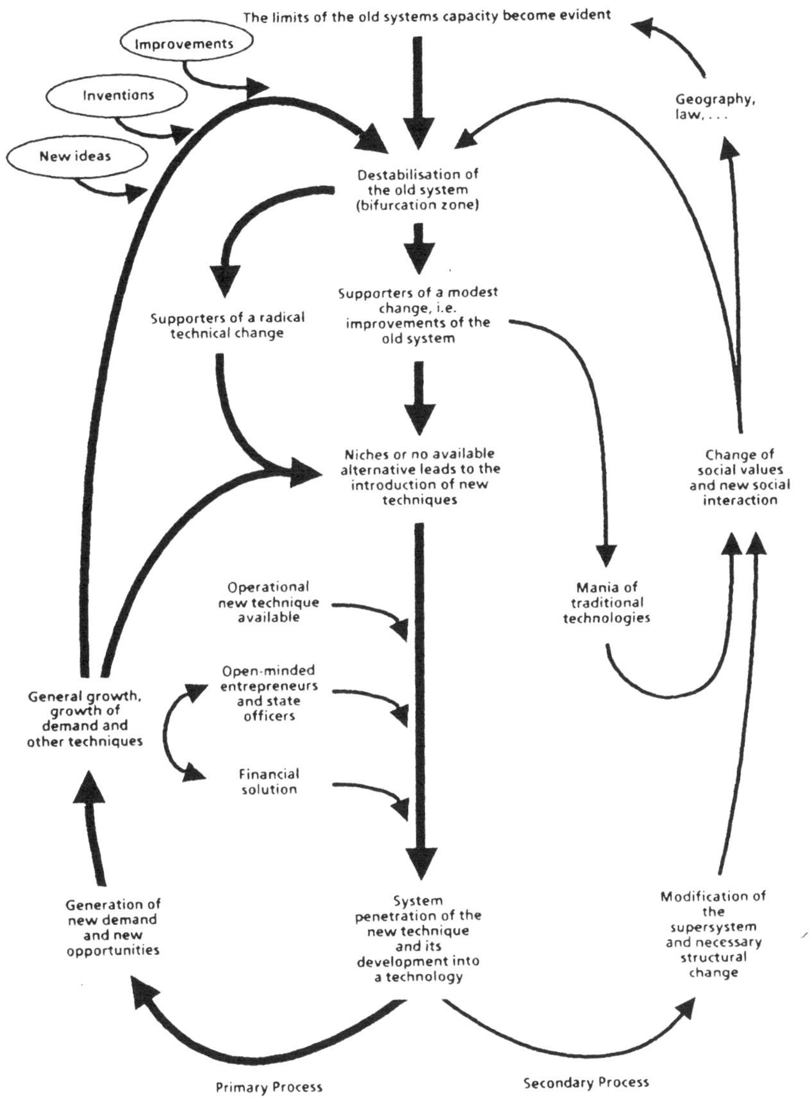

can be represented as a four-step acceleration process: (1) the existing system reaches its capacity constraints; (2) a new technology is at hand; (3) improvements of the old technology and isolated usage of the new technology increase demand for transportation; (4) this additional demand allows the full engagement of the new technology (Figure 6).

Figure 6:
THE DYNAMIC MECHANISM OF SUCCESS

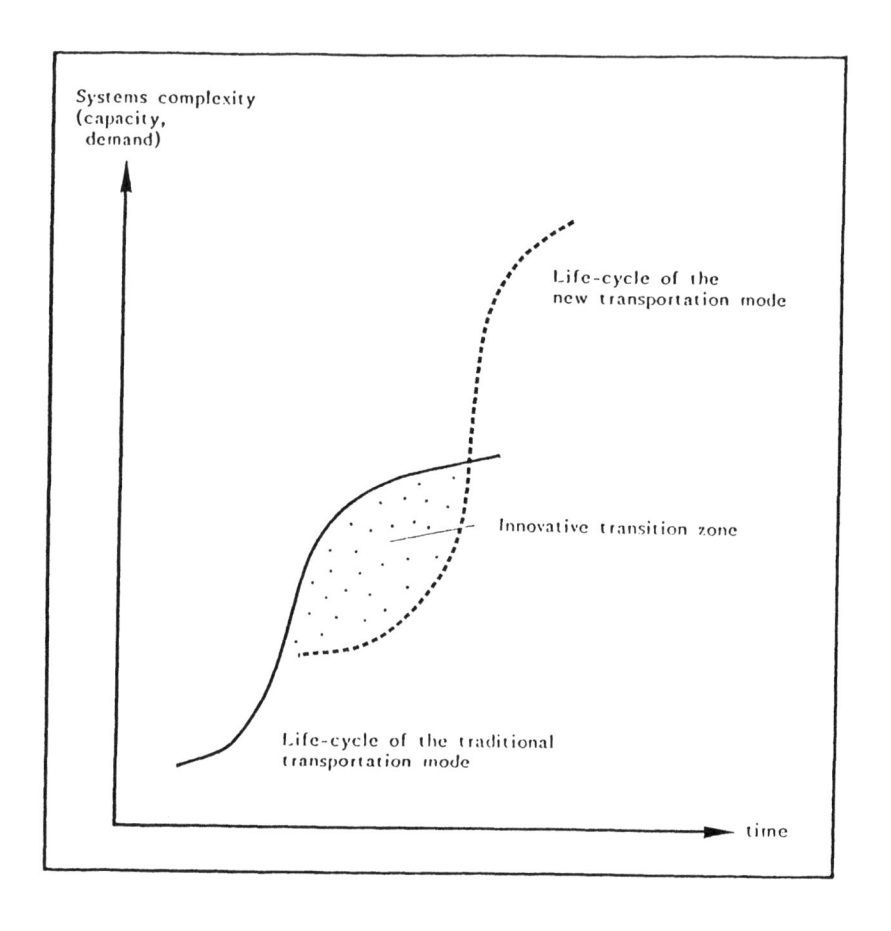

After another 70 years and after the diffusion of the automobile, the railroad network that is used by the Deutsche Bundesbahn today is very much the same as that part of the network that had existed at the end of the demand-oriented construction stage.

The rise and fall of the German railway system suggests some interesting conclusions. The growth of urban agglomerations and conurbations was connected with the building of the railway system. A by-product of this spatial differentiation was the creation of "the rural areas". When the main lines had been built, the search for new investment opportunities for rail products, together with considerations of regional equity, stimulated further extension. Because no better means of transport was then available, the feeder rail network had to fill the gap. This decreased the profitability and efficiency of the whole network, but the transport monopolist of that time could afford it. However, the need for a more flexible and faster low cost solution for areas of low demand was now felt. To provide everybody with a private siding was beyond the abilities of the railway system; the highest network density was reached when 75% of the villages were accessible by rail. Thus a niche opened up for the automobile.

A critical point was reached in the 1920s when the railways came under competitive pressure from road haulage. Caught in a world of railway thinking, the policymakers and their advisers were unaware of the different quality and cost profiles of this newcomer. They also underestimated the structural changes in general transport demand. This miscalculation was portentous because the increasing percentage of high-value goods favored the lorry. The political solution (besides licensing) was to tie the prices of road freight transport to the high tariffs of the railways. This meant that the railways were not compelled to concentrate on the market segments they were especially suited for. On the contrary, these political measures fed the dangerous illusion of being still the general national carrier. The result was inevitable: the high prices enabled the road haulers to challenge the railways on their own ground. The high prices not only shifted the high value goods to the road, but the high profits enabled the road haulers to compete with the railway in the market for low value mass-product transport between industrial and commercial premises where a substantial volume of traffic originates or is discharged. In the 1950s, railway managers were ready to lead their state enterprise like a private business. But the "Rail Act" remained unchanged and forced the railways to continue behaving as a general means of transport. Whereas the road and its transport got nearly everything it wanted from legislators, the railroad did not. The end of the story is well-known.

Notes

1 Similar stages of railway development can be found in: T. Haberer, Geschichte des Eisenbahnwesens. Wien, Pest und Leipzig: A. Bartleben, 1884; G. Riegels, Die Verkehrsgeschichte der deutschen Eisenbahnen. Elberfeld: Baedeker, 1889; H.-P. Schäfer, Verkehr und Raum im Königreich Bayern rechts des Rheins, Habilitationsschrift, Universität Würzburg, Würzburg, 1982 (Manuskript); H. Weigelt, Epochen der Eisenbahngeschichte. Darmstadt: Hestra, 1985.
2 The profitability of this feeder network was doubted by its critics already during its construction.
3 The plans for a unified German customs system, which was in fact demanded by the "Deutsche Bundesakte" (a kind of constitution for the German Confederation) in 1815, failed because of the particularism of the German states. During this time of debate over the German unification Friedrich List became well-known in Germany because of his fight for liberalization and against custom barriers inside Germany.
4 W. Lotz, Verkehrsentwicklung in Deutschland 1800-1900. Leipzig: Teubner, 1906; F. Schnabel, Deutsche Geschichte im 19. Jahrhundert, Vol. 3: Erfahrungswissenschaften und Technik. München: Deutscher Taschenbuch Verlag, 1987 (first 1934); B. Stumpf, Geschichte der deutschen Eisenbahnen. Mainz, Heidelberg: Hüthig und Dreyer, 1961; H.v. Treitschke, Deutsche Geschichte im 19. Jahrhundert. Leipzig: J. W. Hendel, 1927.
5 The Prussian state even financed journeys to England for young engineers so that they could obtain (i.e. spy) the "know-how" of railway building. The intention was economic development by import substitution.
6 Especially the horse-collar and the wagon tongue increased the capacity of horse drawn carriages. Beside this the capacity of the sea going vessels rose from around 100 tons at the turn of the millennium to about 1,000 tons at the end of the 18th century. Improvements in the technology of canal building enabled an extension of the then superior water transport. (F. Braudel, Sozialgeschichte des 15.-18. Jahrhunderts. München: Kindler, 1985).
7 A. Birk, Die Straße. Karlsbad: Adam Kraft, 1933; F. List, Über ein sächsisches Eisenbahnsystem als Grundlage eines allgemeinen deutschen Eisenbahnsystems. Leipzig, 1833; F. List, Das deutsche Eisenbahnsystem als Mittel zur Vervollkommnung der deutschen Industrie, des deutschen Zollvereins und des deutschen Nationalverbandes überhaupt. Stuttgart, Tübingen, 1841. Both published in Friedrich Lists gesammelte Schriften, Ed. by L. Häuser, Th. 1-3. Stuttgart, Tübingen: 1850-1851. The historical importance of List's plan is its design for a region of several

independent states. He foresaw the abilities of this new technology as a tool for the unification of Germany.
8 The procedure of this decision finding is described in great detail by H.P. Schäfer, "Die Entstehung des mainfränkischen Eisenbahn-Netzes", Würzburger Geographische Arbeiten (1979)48.
9 In 1838 the Hofrat G. Muhl from the Duchy of Baden wrote a booklet about the advantages of the railroads. In it, he described a future European railway network that was to bring the world trade from the seaways back to Germany. But for the distance between Köln and Mainz he kept the steamboats. (G. Muhl, Die westeuropäischen Eisenbahnen in ihrer Gegenwart und Zukunft. Karlsruhe: G. Braun'sche Hofbuchhandlung, 1838.)
10 The improvements that were possible become obvious when one considers that the introduction of express carriages alone reduced travel time between Frankfurt/Main and Stuttgart from 40 hours in 1821 to 25 hours in 1822; K. Beyrer, "Das Reisesystem der Postkutsche-Verkehr im 18. und 19. Jahrhundert" in Zug der Zeit - Zeit der Züge: Deutsche Eisenbahnen 1835-1985, Vol. I. Berlin: W.J. Siedler Verlag, 1985, pp. 38-59.
11 W. Klee, Preußische Eisenbahngeschichte. Stuttgart, Berlin, Köln, Mainz: Kohlhammer, 1982.
12 As these railways were copies of the old English wagon ways with flat bar iron on wooden beams and with additional flange wheels they cannot be regarded as the first real (i.e. modern) railways (Weigelt 1985, op. cit., p. 25).
13 Friedrich Harkort's brother Gustav was involved in the building of the Dresden-Leipzig railway and was its first president for 30 years.
14 Schäfer (1982, op. cit, p. 359) cites a list of 17 railway projects that was published by the "Central Verein für Eisenbahnen" in Kassel, 1835.
15 There were others who designed railway plans for Germany at that time as well, e.g. Weigelt (1985, op. cit.,) shows a plan by Grothe from 1834, but these men did not find any publicity at all.
16 See: Schnabel 1987, op. cit., p. 330-393; W.O. Henderson, Friedrich List. London: Frank Cass, 1983.
17 The underdevelopment of German industry and therefore the necessity for all German railway companies to buy their first locomotives in England was the simple reason for the uniform use of Stephenson's 1435 mm gauge. In England it had already become by far the most common one.
18 The first German railway committees usually asked British experts for judgments about their plans. For the first German railway Paul Camille Denise had planned a gauge of 1,420 mm and already started to build it. The necessity to use an English locomotive forced him to change it. The British superiority went so far that this first German steam-locomotive was also put together by an Englishman who later on even drove the train and became a very respected citizen of the city of Nürnberg.
19 Riegels 1889, op. cit., p. 2.
20 Paul Camille Denise had wisely planned low long-term operating costs for this railway. That he could realize his aim, with the exception of the gauge (see Note 18), and did not copy the more advanced example of the Liverpool - Manchester railway, was one of the reasons for this railway's economic success.
21 The weakness of German industry could be cited as a fourth factor. Several projects got delayed because domestic industrial products could not be delivered in time or broke during construction.

22 J.C. Bongaerts, "Financing Railways in the German States 1840-1860 - A preliminary view", The Journal of European Economic History, 14(Fall 1985)2, pp. 331-345.
23 There were two other railways longer than 100 km in Germany before the Cologne-Minden railway. The "Linz-Budweis railway" connected the rivers Moldau (Elbe) and Danube already in 1832 with a length of 131 km. It is not considered here because horses were used to pull the wagons. Another reason is that its construction did not even allow steam engines. So already in 1859, parts of it were removed and in 1874 the complete line was closed (G. Kleinhanns, "Die Linz-Budweiser-Bahn - Ausgangspunkt des internationalen Schienenverkehrs", Oberösterreichische Heimatblätter 36(1982)3/4, pp. 250-259). Also about 1840 Dresden was connected with Leipzig and Leipzig with Magdeburg. But this 230 km link was not designed as one railway.
24 An inherited belief from the middle-age road policy was to keep through traffic as long as possible on one's own territory, mainly for financial reasons and for the sake of wheelwrights, innkeepers and postmasters. Railheads were not only built because of the multi-state system. Another reason was the belief that changing trains and/or shifting cargo were good for the local economy.
25 Riegels 1889, op. cit., p. 27.
26 E. Jochem, "Hilfen und Irrtümer beim Rückgriff des Prognostikers auf die Vergangenheit", VDI-Technikgeschichte, 51(1984)4, pp. 263-275.
27 During the first two decades the railways did not intend to compete with water transport. Therefore in Northern Germany with its south-north going rivers mostly East-West connections were built. Railways that connected two cities at the same river tried to avoid the river valley. F. Voigt, Verkehr, Vol. II: Die Entwicklung des Verkehrssystems. Berlin: Duncker und Humblot, 1965.
28 W. Klee 1982, op. cit., p. 96.
29 The reduction in travel time between major destinations was achieved to a much greater extent by shortening the waiting time at the stations on the way or even by passing them, than by rising the speed of the trains (Voigt 1965, op. cit.).
30 F. Voigt, "Die Einwirkungen der Verkehrsmittel auf die wirtschaftliche Struktur des Raumes - dargestellt am Beispiel Nordbayerns" in Die Nürnberger Hochschule im fränkischen Raum 1955. Nürnberg: Glock und Lutz, 1955. F. Voigt, Die gestaltende Kraft der Verkehrsmittel in wirtschaftlichen Wachstumsprozessen. Untersuchung der langfristigen Auswirkungen von Eisenbahn und Kraftwagen in einem Wirtschaftsraum ohne besondere Standortvorteile. Bielefeld: Kirschbaum, 1959. H. Weigelt, "Bayerische Nebenbahnen im Strukturwandel der Flächenbedienung", Die Bundesbahn, 60(1984), pp. 153-162.
31 The impacts of the railway on all aspects of the society are shown by: F. Schnabel 1987, op. cit.; W. Schivelbusch, Geschichte der Eisenbahnreise. Zur Industrialisierung von Raum und Zeit. München, Wien: Hanser, 1977; F. Voigt, Die volkswirtschaftliche Bedeutung des Verkehrssystems. Berlin: Duncker und Humblot, 1960.
32 O. Lang, "Die Eisenbahn in der Weimarer Zeit" in Zug der Zeit - Zeit der Züge: Deutsche Eisenbahnen 1835-1985, Vol. 2. Berlin: W.J. Siedler Verlag, 1985, pp. 654-660.

CHAPTER 5
LOOKING FOR THE BOUNDARIES OF TECHNOLOGICAL DETERMINISM:
A BRIEF HISTORY OF THE U.S. TELEPHONE SYSTEM[*]

Louis Galambos

1 The task at hand

The task to which the following pages are devoted is three-fold. First, I am to examine the development of the U.S. telephone system, ascertaining the extent to which technology and other factors - including politics - shaped its structure. Second, I am to determine whether and to what effect the telephone network became or contained a technological system or systems which acquired the type of socio-economic momentum that Thomas P. Hughes found in electrical power systems. Third, I am to outline the strategies of the major actors - including the state - and facilitate comparative analysis by specifying the dominant modes of telephone utilization. Since the system in this case is extremely large (including one firm which was until recently the largest private business corporation in the world), and since the American telephone network and the Bell System have been the subject of many books and articles, I must skim like a hovercraft over the surface of an immense sea of information. Fortunately, I have found help in developing a suitable perspective. In the recent antitrust suit, U.S. v. AT&T, the defendant contended that its structure and behavior were for the most part determined by technological factors and by the company's efforts to remain efficient and innovative within those technological parameters. Hence I can use the so-called "Gold Book" in which AT&T set forth its contentions and proof as a buoy marking one side of the interpretive channel I use.[1] I can employ Gerald Brock's interesting book on The Telecommunications Industry to mark the opposite side of that channel, providing the black buoy, as it were.[2] Brock develops a variant on neo-classical economics and critiques the Bell System and the industry for deviating from a competitive model, for exploiting regulatory controls to protect their interests, and thus for being less efficient and innovative than they should have been. Brock gives little heed to technology as a causal force. In my own summary analysis, I steer clear between those two buoys.

2 The birth of the Bell System

In the first phase of telephone development in the U.S., from the initial invention in 1876 until 1878, when the exchange technology was developed, there were limited economies of scale and scope and none of system. Transmission was from point to point, with each telephone pair constituting a separate network. Given this situation, you can almost excuse the Western Union Company's lack of vision when it turned down the opportunity to buy the Bell patent.[3] Later, the telegraph firm moved into telephony briefly, but then it settled its dispute with the fledging Bell interests by withdrawing from the industry entirely. Western Union was at the time the largest corporation in the United States and the dominant firm in telegraphy. Its officers apparently envisioned the telephone as nothing more than a means of extending Western's nationwide information system by distributing on a local level the messages carried over its long-distance wires. Even the exchange technology, which first introduced economies of system, did not alter a Western Union strategy framed in terms of its primary market, that is the business customer. Telegraphy gave business customers a written record of their transactions, something the telephone could not do. Add to that the machinations of robber baron Jay Gould and you have the rationale for Western Union's 1881 settlement with the Bell interests.[4]

The settlement came none too soon for Bell, because the company's financial resources had been drastically strained by its efforts to compete with the giant telegraph company. Each had raced to establish exchanges in the major urban markets, and as a result, by 1881 there were 71,387 telephones in service in the U.S. But bear in mind that during these early years the patents were about all that Bell had going for it. This explains Bell's decisions to lease (not to sell) telephones to users and to produce all of the telephones itself. Only in that way could it successfully fight off patent infringement, a goal that it achieved from 1881 to 1894/95, when its patents expired.[5] Since Bell did not sell the phones, however, it needed more capital than would otherwise have been the case. The exchange technology was particularly expensive. Lacking the resources - capital and entrepreneurship, as well as administrative ability - needed to promote telephone expansion across a very large and diverse nation, Bell licensed local agents to establish phone companies that could raise local capital and build the urban exchanges. The local entrepreneurs in this highly decentralized operation at first also provided technological inputs, fostering rapid improvement in switching equipment.[6]

In this phase of development, the two dominant influences on the network were thus the patent law and the Bell organization's lack of capital and personnel to promote growth on a national scale. Technological considerations alone dictated a series of separate urban companies, each with a monopoly in its local market. There was no technological rationale for having these separate firms joined in a single business system. While Western Electric, after 1882 the primary supplier of phones and equipment to the Bell companies, realized economies of scale and doubtlessly eliminated some transaction costs in purchasing, it seems highly unlikely that this alone would have caused a unified system to develop in the absence of the patent law. The result of that law and Bell's lack of resources was a telephone system consisting of a single private firm which operated in a highly decentralized manner, employing local agents to promote development.[7]

The decentralized structure had long term consequences for the network, including its technology. As the local phone companies evolved, each initially employed its own variant on the basic technology, using similar but slightly different equipment and wires. At first these differences did not matter very much. But once Bell began to push long-distance transmission, the technical and financial problems of connecting the exchanges became acute. This set in motion a twenty-year process in which Bell gradually and hesitantly developed first switching and then transmission technology and imposed them on the network. Protected by the patent monopoly and better able after 1881 to build up capital resources, Bell acquired controlling interests in the local companies, in part to ensure that technical standardization and coordination could be achieved.[8]

While these goals were being pursued, Bell management emphasized high profits over expansion of service - just what you would expect from a monopoly.[9] As a result, most of its customers were businesses, and this was particularly true of the emerging long-distance service. The Bell System - as it came to be known - continued to concentrate its expansion in the most lucrative urban markets, largely ignoring potential rural and small town customers. By 1895, when the patents expired, there were 309,502 telephones operating in the United States and all of them were owned by American Bell, a firm characterized by its own attorney as the most hated patent monopoly in the nation.[10] To that date neither the state nor competition had played a positive role in the development of this evolving network. Bell had achieved economies of system and of scale, had used its protected position to reap unusually high profits, and had promoted steady, selective growth. Bell management had also emphasized technical innovation, and as the technology for long-distance service improved, the economically optimal size of firm had grown; economies of system and scale by 1895 favored an industry comprised of several large regional firms, somewhat along the lines of the current regional holding companies but combining long-distance and local service. Instead, the industry was monopolized by a single business, American Bell whose primary objective had been neither technical innovation nor operating efficiency: management's major goal through 1895 had been market control and the profits it yielded.[11]

3 The effects of competition

After the mid-1890s, competition drastically altered the network's pattern of growth, and in the years that followed, the local, state, and federal governments became active participants shaping network development. As the patents expired and the courts rejected Bell's efforts to extend their life, competitors rushed into the industry, forcing prices down, speeding diffusion of service, and ultimately precipitating a decisive managerial and structural transformation in the Bell System. Between 1895 and 1907, the new competitors, called independents, installed almost three million new telephones and the Bell System about 2.7 million new sets. By the latter date, the industry was evenly divided between Bell and the independents, although Bell controlled virtually all of the long-distance lines that linked the various urban exchanges. Bell refused interconnection to the competing independents so the nation had, in effect, two telephone systems. One, the Bell System, was technically integrated on a regional and inter-regional basis (to be specific, you could call Chicago from New York but not Denver or San Francisco); Bell was national in scope and stressed a broad range of high-quality, high-cost services. The other part of the industry consisted of a wide variety of separate, local systems which generally charged low prices for a limited range of services that varied significantly in quality.[12] Some of these systems were tiny farmer cooperatives. The largest ones were urban companies going head to head with the Bell System in struggles for the control of city markets.

The competitive era after 1895 greatly accelerated telephone diffusion in the United States and changed the institutional setting of telephony in two major ways. It was accompanied by government intervention in a significant and positive manner, first at the local, then the state, and finally the federal levels. Municipal governments had always been involved in certifying and authorizing rights-of-way for their local telephone systems, but their choices had been limited before 1895 to Bell service or no service at all. After 1895 they had choices, and they also had new groups of local entrepreneurs calling for municipal support in a fight against the trust, a popular cause in turn-of-the-century America.[13] State governments too were inspired to establish regulatory commissions or to use existing regulatory bodies - usually railroad commissions - to bring their state telephone networks under a measure of public control. By the end of 1907, twelve states had passed specific statutes embracing telephone regulation and most of these adopted the style of rate-of-return regulation that was in those years gradually being imposed on the nation's railroads.

Not until 1910, however, did the federal government become involved in telephony. National control emerged slowly because traditionally in the United States regulatory functions had been reserved to the states. The first major break with that tradition came in 1887, when Congress established the Interstate Commerce Commission (ICC) to control a railroad network that had long before outgrown the states and their political powers. Even then, it was 1910 before such regulation actually became effective. The Mann-Elkins Act of that year strengthened the ICC and extended its common carrier authority to encompass interstate telephone companies. The Commission began its work by endeavoring to establish a standard accounting system and by answering complaints, leaving most regulatory functions in the hands of the states. It would be 1934, before the federal government through the newly formed Federal Communications Commission would impose thorough-going regulation on interstate telephone service.[14] Federal government ownership and operation was not really considered to be a viable option in the United States at the beginning of the century, in part because the administrative federal state had just begun to take shape and in part because public attitudes still favored private ownership. As late as 1900 the national government lacked the expertise, capital, and authority to run the telephone system as a state-owned monopoly. Some reformers called for this solution, and there were many advocates of municipal ownership of local exchanges; but nationalization was never really a likely alternative. Regulation was.

In addition to fomenting regulation, the competitive era so changed the industry that it forced a dramatic shift in management and in corporate strategy in the Bell System. In 1900 the American Telephone and Telegraph Company - established in 1885 as the long-distance subsidiary - became the central holding company of the Bell System, and as AT&T expanded in an effort to forestall competition, the firm ran into financial difficulties.[15] It borrowed heavily to finance growth. By 1907 AT&T had overtaxed its resources and was unable to float the bonds it needed to finance further growth. At this point, New York investors - in particular banker J.P. Morgan and his associates - supplanted AT&T's Boston financiers and selected Theodore Vail to head AT&T.[16] It was Vail who would revamp, restructure, and reorient the firm, creating the modern Bell System and a national telephone network that would for many decades be the envy of the world.

4 System building: The development of a large corporation

Under Vail's leadership, AT&T would again dominate the industry and would strike a new balance between efficiency, innovation, and market control. The Bell enterprise would for the first time become the sort of technological system that Thomas P. Hughes has identified as a central concern of this conference. Operations were reorganized throughout the System, first in the long distance service (Long Lines) and then among the operating companies: the basic structure employed was a three column functional organization framed in terms of the plant, traffic, and commercial aspects of the business. The common mode of organization facilitated communication and helped AT&T eliminate redundancy in the System. Under Vail, AT&T accelerated the processes of technical standardization and of internalization of technical innovation. The Bell System consolidated its various research and development centers (ultimately in Bell Labs, 1925) and strengthened its ties to major sources of scientific and engineering progress in the United States (and abroad).[17] As this process of integration began to take hold, the System developed momentum: in this case what that means is that there emerged a powerful corporate culture that stressed operational efficiency and ongoing technical improvements; there was a social system of rewards and punishments that reinforced those values; and there was a corporate structure that emphasized functional expertise in achieving system-wide goals of improved performance. Vail condensed those goals into a slogan that would epitomize Bell strategy for more than half a century: "One system, one policy, universal service".[18]

These goals could not have been achieved, of course, if AT&T had not worked out a new accommodation with public authorities in the states and the national government. Under Vail's leadership, the Bell System accepted state-level rate-of-return regulation and cooperated with the regulators, the most important of whom were then at the state level in the American federated style of government. Most U.S. businessmen during these years fought against the emergence of a regulatory system, but Vail recognized that AT&T had to trade off some of its autonomy in order to maintain a dominant position in the industry. Cooperation with public authority became one of the traditions of the modern Bell System. Indeed, the crucial element determining the degree of decentralization of the System after 1907 was the desire to let the operating companies retain the authority they needed to deal with their separate state regulatory commissions. Telephone service was very much a local matter, and telephone politics, not telephone technology, called for the retention of a structure that was technologically centralized but economically and politically decentralized. After the network was technically integrated on a national level, following the development by 1915 of transcontinental transmission, AT&T enjoyed significant economies of system - as well as scale.

The technology dictated an integrated, centralized firm of the sort developed by most large industrial producers in these years. But the politics of regulation called for compromise on this and other fronts.[19]

Vail also compromised with federal authority in the Kingsbury Commitment of 1913. Under Vail's leadership, AT&T had begun to promote a vigorous policy of acquiring some independent telephone companies and encouraging other, non-competing independents (usually in semi-urban or rural settings) to interconnect with the Bell System. In 1908 AT&T also absorbed the Western Union Company (which had earlier spurned an offer to buy the Bell patent rights for $100,000). Between 1907 and 1913 the number of Bell telephones increased from about three million to well over five million; the number of non-connecting independent phones dropped to about one and a half million. There were by 1913 around 2.8 million independent phones connected to Bell's network.[20] The Bell System used subterfuge and highly questionable tactics in the drive towards monopoly, and some of the independents responded with antitrust suits against AT&T.[21]

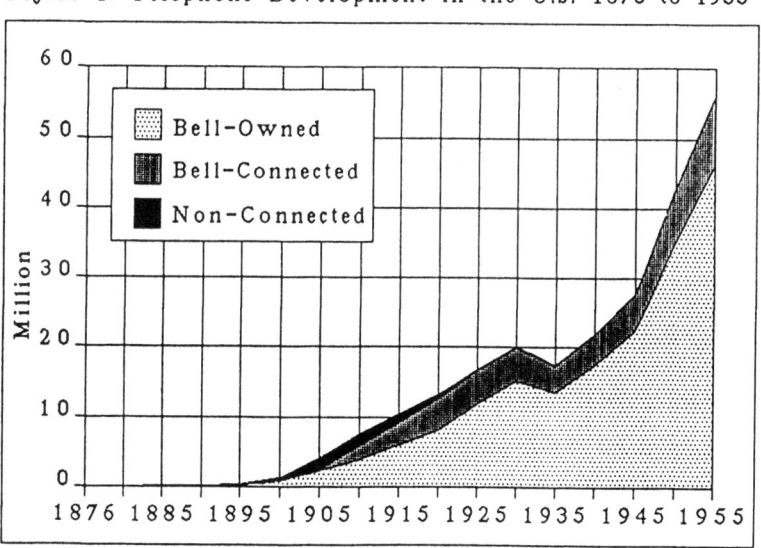

Figure 1: Telephone Development in the U.S. 1876 to 1955

By 1913 antitrust had become a real threat to combines like AT&T, especially after the 1911 Supreme Court decisions that broke up the Standard Oil and American Tobacco trusts. In 1913 when the Department of Justice expressed concern about AT&T's latest acquisitions and filed an antitrust suit, the firm agreed to take the following actions: AT&T sold its controlling share of the Western Union telegraph company; it agreed for the first time to interconnection with the competing independents, allowing them to use AT&T's long-distance lines. The letter that AT&T Vice-President Nathan C. Kingsbury sent to the U.S. Attorney General also said that the company would acquire no additional competing independents without government approval. As the Commitment was interpreted, Bell was allowed to trade stations with independents so that each could round out their systems, and by 1921 the Bell System had 8.7 million phones in place; among the independents 4.6 million phones were integrated into the Bell network and less than half a million were non-connecting. By this time about 35% of American households had phone service.

Interconnection and the perfection of long-distance service created a truly national system, and in the years that followed, the concept of the network acquired a powerful mystique among Bell System managers and workers. It was embodied in systems engineering (which had its origins at Bell Labs),[22] reflected in nationwide rate averaging, and maintained through a system of executive advancement that regularly brought managers up through a series of jobs in operations before they graduated to AT&T's national headquarters - the so-called General Departments. From the General Departments they monitored the performance of the System's horizontal (the operating companies) and vertical (Bell Labs and Western Electric) components. While all of these components were linked to AT&T by ownership of stock (in varying degrees), control from the center was relatively light-handed, in large part because there was so little internal disagreement about the System's goals and values.

The Bell System grew faster than the independents and also continued - under governmental supervision - to acquire telephones from and minority ownership positions in some of the independent companies. On the eve of America's entry into World War II, the System owned 83% of the telephones in use and almost all of the long-distance lines. By this time, plain old telephone service ("POTS" in the industry) to residential customers accounted for two-thirds of the nation's telephones, but businesses still made most of the long-distance calls. Western Electric supplied about 90% of the Bell System market for telephone equipment.

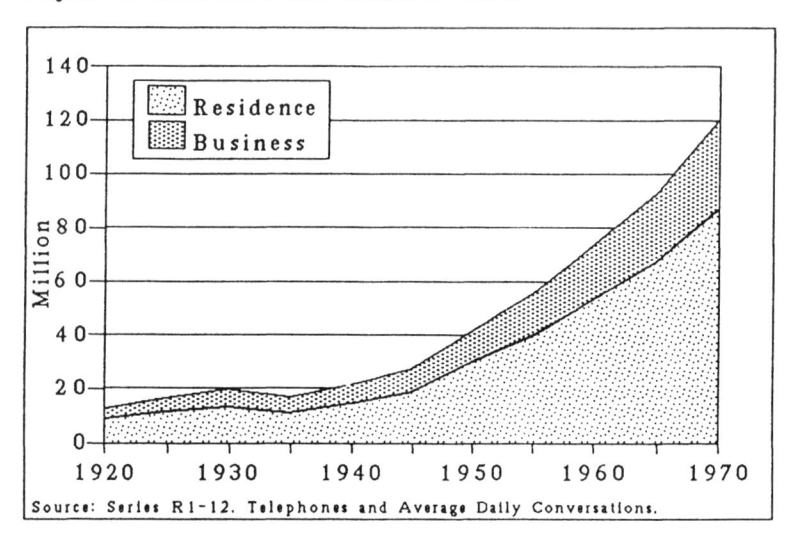

Figure 2: Residence and Business Users

[Figure: Stacked area chart showing Residence and Business telephone users in Millions from 1920 to 1970, rising from near 0 to approximately 130 million. Source: Series R1-12. Telephones and Average Daily Conversations.]

The Bell System and the independents were, however, extensively regulated, and were forced by rate-of-return controls to accept relatively modest increases in prices (called rates). This situation was made tolerable over the long term by significant increases in productivity, but nevertheless, Bell System stockholders did not reap significant capital gains. Instead they received moderate, regular dividends. Indeed, the hallmark of the modern U.S. phone system was steady progress in a setting that involved minimal risk. Since the independents interconnected with the Bell System and were protected by public policy, they had nothing to fear from AT&T. Since entry to the industry was controlled by state and federal authorities, AT&T was subject to risk on only three fronts: managerial failure to adapt to fluctuations in the economic and business setting; potential competition from new technologies like radio; and the threat of some form of governmental action, either by federal officials or by state-level regulators.

AT&T's best defense against all of these threats was to remain technologically and organizationally progressive and to work toward the goal of "universal service". By 1970 with over 120,000 phones in place in the United States and over ninety percent of the households with telephone service, that goal had been achieved. Meanwhile, so much progress had been made in long-distance technology that AT&T and federal and state regulators had implemented a plan (called "separations") which in effect paid a large and growing subsidy from long-distance to local service.[23] In these years of the so-called "American Century", Bell Labs was widely acknowledged to be the premier industrial lab in the United States, if not the world. Neither the government nor independent consultants could find any serious problems with Western Electric's performance.[24] Indeed, Western Electric and the entire Bell System were a theoretical contradiction: a bureaucratized, near monopoly that was efficient and innovative.

Still, the position of a near monopoly, the largest firm in the world, a private company controlling a vital mode of communications, was tenuous vis-à-vis the federal government. Efficiency was generally no defense against an antitrust suit, and that was particularly true if the company was taking actions that in the view of the Department of Justice were aimed at maintaining its monopoly. In the 1930s in the midst of the Great Depression, anti-business sentiment had mounted and had resulted, as we have seen, in the formation of a new, more powerful regulatory commission for telecommunications, the Federal Communications Commission. The anti-business, regulatory movement that produced the FCC eventually spilled over into a 1949 antitrust suit that was finally settled in 1956, with an important consent decree. Under the decree accepted by the Eisenhower Administration, AT&T kept all of its component parts. AT&T was, however, barred from entering unregulated lines of business and was required to make all of its technology available to others on reasonable terms. Since the government was dealing with a firm that had invented the transistor (1947), this latter requirement represented a significant concession on the part of the company. Nevertheless, the 1956 consent decree favored the Bell System. It was in part a product of an administration friendly to big business. But it was also a result of the System's long-run success in achieving Vail's goals and its excellent performance in contributing to the national defense effort during and after World War II.

5 The monolith is challenged

In the aftermath of this settlement, the future for telecommunications looked bright and stable: the federal government (with some exceptions in the Justice Department) was happy with the 1956 consent decree; the state regulators were satiated by the subsidies flowing to local service; the independents were protected and their profits shielded from competition; universal service was achieved; the public was pleased with its phone service; the Bell System was triumphant. But soon this public-private equilibrium began to break down, and the result ultimately would be a restructured telecommunications industry which would attempt to replace the internal momentum heretofore generated by the Bell System with market forces. The events and historical processes that unsettled the seemingly productive relationships that existed during the early 1960s between private telecommunications firms and public power clearly demonstrated how tightly interwoven in the modern Bell System were the elements that fostered efficiency, innovation, political accommodation, and market control.

To some extent the forces of change were technological and to a considerable extent the shifting technology was produced by the Bell System itself. One important development was microwave transmission which from the late 1940s on made long-distance service less of a natural monopoly.[25] A pricing structure that featured nationwide averaging and subsidies flowing from long-distance to local service created unusual opportunities for profit if only entrepreneurs could penetrate the network and take advantage of the new microwave technology. Another significant technological change was the shift from electro-mechanical to electronic switching, a development which made it difficult for a single company - even one that had been as successful as Western Electric had been - to remain at the forefront of all phases of telecommunications technology. Other manufacturers had been gaining experience in the electronic technologies since World War II and were well positioned to move into telephone equipment markets. A third and closely related development was in computer technology, which generated new demands in data transmission. This phenomenon put pressure on the Bell System to respond in new ways for which the existing network was not designed to the variegated needs of individual business customers.[26] Later, computers also began to generate new opportunities that AT&T's management found awkward or impossible to exploit because of the 1956 Consent Decree. But of course these technological changes only created theoretical or potential opportunities for other firms so long as entry into telecommunications markets was prevented by federal and state authorities.

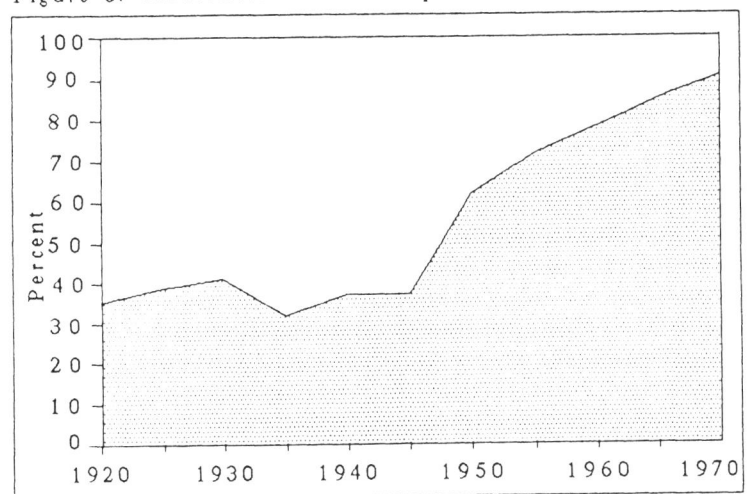

Figure 3: Households with Telephones 1920 to 1970

Entry was the crux of the matter and the decisive shift took place on this front. During the 1960s, the Federal Communications Commission and the courts began gradually, hesitantly, and at times inadvertently, to open the doors to entry. The available evidence indicates that the FCC did not seek to restructure the industry to any significant degree. But once the process began, it was difficult to arrest because the entrepreneurs who took advantage of these new opportunities became active, aggressive, and frequently very successful participants in the political and legal developments that ultimately brought down the Bell System.[27]

AT&T contributed to its own downfall. Dismayed by the changes that were taking place and uncertain how to respond, AT&T management allowed a sense of drift to develop within the Bell System in the late 1960s and early 1970s. Earnings sagged, as did service. This was the situation John deButts inherited when he took over AT&T's chairmanship in 1972. A vigorous leader, deButts mustered the Bell System's considerable resources behind the Vail banner. He refused to compromise with federal authority and attempted to preserve intact AT&T's dominant position in all phases of the industry, a position that he felt was essential to the preservation of the network and universal service.[28] Within the System, deButts' firm leadership restored morale and pushed service quality and earnings to record high levels. But that policy left the Bell System vulnerable both to private and to public antitrust suits. In effect, AT&T improved its efficiency and tried to preserve its market control, but the price was high: the firm lost control of its political environment in an unsuccessful effort to breathe life into the Vail tradition.

On the heels of regulatory change and private antitrust suits came a federal antitrust suit brought against AT&T by the Department of Justice in 1974. The federal suit dragged on through the rest of the decade, while Congress deliberated over but failed to pass a new telecommunications law. Finally, deButts' successor, Charles L. Brown, was forced in 1982 to accept most of the government's terms and divest the local operating companies. Brown chose that course rather than continue the legal fight (leaving the company's future in doubt) and risk the chance that AT&T would lose its vertical components, Western Electric and Bell Labs, when Judge Harold Greene rendered his decision in the antitrust case. Without Western and the Labs, AT&T could not compete effectively, Brown thought, in the global struggle he foresaw for Information Age markets.[29]

The divestiture agreement of 1982 left U.S. telecommunications structured along the following lines: seven separate and very large regional holding companies controlled local services; AT&T was still the dominant firm in long distance but it had several competitors, the largest of which was (and still is) MCI. AT&T kept its manufacturing and research facilities, and it has attempted to transform itself into a competitive firm while maintaining the high level of innovation that long characterized the regulated Bell System. AT&T has had to change in many ways. Even before the settlement, equipment markets had become extremely competitive; foreign and domestic producers had cut deeply into the share of the market that AT&T had once controlled. That process has continued. Equipment markets are completely deregulated; in long distance, AT&T is still subject to extensive rate-of-return regulation, but its competitors are free of government controls. Most of the local phone companies are still regulated in the traditional manner.[30]

In the restructured telecommunications industry, managers and public officials have had to develop new strategies. Firms can no longer balance efficiency, innovation, and control of the market and political contexts in the way that Vail did. Time horizons have had to be shortened. Meticulous planning for a national system gave way to marketing. As subsidies were reduced and competition mounted, POTS became more expensive and long-distance cheaper. Even the regional holding companies have had to behave in new ways: they can no longer control the institutions that perform the research, development, and manufacturing functions; meanwhile, some large customers have begun to bypass the local companies entirely by putting in their own private lines connecting to the national network. Regulators too have been forced to adjust to a newly competitive industry. The loss of subsidies has forced state commissions to raise rates. The FCC's role has changed, in part because of deregulation and in part because Judge Harold Greene, who oversees the antitrust settlement, has in effect become the industry's chief regulator on many questions of structure and performance.[31]

The new structure of the industry was largely a product of political, not technological, factors. Only in equipment could the outcome be said to follow the lines dictated by largely technical forces - that is, by the need to have a broader range of producers to maintain the level of innovation required in this fast-changing industry. The regional companies were and are still today barred by the consent decree from manufacturing and long distance. They would like to enter both of these markets if they could. Meanwhile AT&T is still going through its shakedown cruise as a competitive firm. It has made considerable progress in dealing with the needs of its business customers, but it has been difficult to transform the organization from a technologically oriented to a market oriented posture. In the emerging era of fiber optics, AT&T could become once more a natural monopoly in long-distance service. If so, it is unlikely even to try to drive any of its major competitors to the wall (sustaining them as it long did Western Union). Careful to avoid another adverse government reaction, AT&T seems likely to share the market with its competitors, although on terms favorable to AT&T (which today has about 80% of the market).

So the history of U.S. telecommunications shows at every stage how a technological system was and is shaped by its political context. Early telephony by the patent laws. The Vail system by regulation and antitrust. The more recent industry by regulatory change, antitrust, and now the Federal Judge who is overseeing the settlement. Only in the middle phase, the years of the Vail settlement, did a full scale technological system, à la Hughes, evolve in this industry; this system was unique in many regards, especially in the extent to which it remained an innovative institution. From the early 1900s to the mid-1960s, the balance Vail had struck between efficiency, innovation, and control of the firm's market and political contexts served AT&T, the Bell System, and the United States well. When, however, a changing environment called for a new balance, a new set of compromises and trade-offs, AT&T's management fought so hard to preserve the traditional strategy that it lost control of its political setting. Within a decade political forces had destroyed the integrated national network as the United States looked to market forces for the technical momentum that the Bell System had generated for most of the twentieth century.

Notes

* I appreciate the assistance I received from Robert Lewis and Alan Gardner at the AT&T Archive and from the History Department - especially Betty Whildin and Susan Mabie - at the Johns Hopkins University.
1 United States of America v. American Telephone and Telegraph Company, et al., Civil Action No. 74-1698, Defendants' Third Statement of Contentions and Proof.
2 Gerald W. Brock, The Telecommunications Industry: The Dynamics of Market Structure. Cambridge: Harvard University Press, 1982.
3 Robert W. Garnet, The Telephone Enterprise: The Evolution of the Bell System's

Horizontal Structure 1876-1909. Baltimore/Md.: Johns Hopkins University Press, 1985, p. 11.
4   Ibid., pp. 29-44. George D. Smith, The Anatomy of a Business Strategy: Bell, Western Electric and the Origins of the American Telephone Industry. Baltimore/Md.: Johns Hopkins University Press, 1985, pp. 35-80. See also Rosario Joseph Tosiello, The Birth and Early Years of the Bell Telephone System: 1876-1880, Unpublished Ph.D. Dissertation, Boston University, 1971, pp. 452-491.
5   John Brooks, Telephone: The First Hundred Years. New York: Harper & Row, 1975, pp. 76-81. Garnet 1985, op. cit., p. 15.
6   Smith 1985, op. cit., pp. 63-64.
7   Two other factors shaping this outcome were the lack of vision on the part of Western Union's management and the inability of the United States government to foster a development such as telephony. In the late nineteenth century the American government had very little administrative capacity and few resources. State-level experiences in the early part of the century with government operation of canals had been discouraging, and American attitudes and values favored private entrepreneurship (although often with government support) over public efforts.
8   This process of technical consolidation is one of the central topics of Robert W. Garnet's The Telephone Enterprise.
9   During the seventeen years of the patent monopoly, the average annual return on investment was about 46%, according to Brock 1982, op. cit., pp. 107-108.
10 James Storrow, American Bell's counsel, wrote the firm's president in 1891: "The Bell Company has had a monopoly more profitable and more controlling - and more generally hated - than any ever given by any patent". As quoted in Brooks 1975, op. cit., pp. 102-103.
11 The relationships between efficiency, innovation, and market control are discussed at greater length in Louis Galambos, "What Have CEO's Been Doing", Journal of Economic History (forthcoming), and in Louis Galambos and Joseph Pratt, The Rise of the Corporate Commonwealth: U.S. Business and Public Policy in the Twentieth Century. New York: Basic Books. Bear in mind that until the development of the common battery system following 1895, many subscribers had to have two telephones: one for local calls and the other for long distance. Hence there was no network in the modern sense of that word.
12 For an excellent discussion of these struggles see Kenneth Lipartito, The Telephone in the South: A Comparative Analysis, 1877-1920, Dissertation, Baltimore/Md.: Johns Hopkins University, 1987.
13 On public attitudes toward the large corporation see Louis Galambos, The Public Image of Big Business in America, 1880-1940: A Quantitative Study in Social Change. Baltimore/Md.: Johns Hopkins University Press, 1975, especially pp. 47-114.
14 The FCC's regulation was informal until the mid-1960s, but roughly the results were the same as formal, state-level, rate-of-return regulation. Bell management was very concerned during the early 1900s about the possibility that urban governments would take over some of the local exchanges, preventing Bell from developing the sort of integrated network the company was trying to create.
15 The Bell System's telephone plant had grown 6.3% and 8.8% in 1891 and 1892. In the early twentieth century, the comparable figures were: 1900=14.2%, 1901=17.2%, 1902=18.0%, 1903=13.8%, 1904=11.2%, 1905=16.3%, 1906=22.3%. Garnet, The Telephone Enterprise, pp. 164-165.
16 Vail had previously served as president of AT&T, when it was a long-distance

subsidiary, and of Metropolitan Telephone and Telegraph Company, New York. He had left active management in the Bell System in 1889.

17 See, for instance, Leonard S. Reich, The Making of American Industrial Research: Science and Business at GE and Bell, 1876-1926. New York: Cambridge University Press, 1985, pp. 142-217. See also Neil H. Wasserman, From Invention to Innovation: Long-Distance Telephone Transmission at the turn of the Century. Baltimore/Md.: Johns Hopkins University Press, 1985. As Wasserman makes clear, the Bell System at the turn of the century was beginning to develop a higher level of scientific expertise and to establish links to the centers of scientific and engineering progress outside the firm. But Bell still lost the race to patent the crucial loading-coil invention, and it lost to a lone inventor. After 1907 the company made a concerted and rather successful effort to ensure that this would not happen again. Also useful is M.D. Fagen (ed.), A History of Engineering and Science in the Bell System: The Early Years 1875-1925. Bell Telephone Laboratories, 1975.

18 See AT&T's 1909 annual report.

19 To some extent decentralization could have been due to the great size and complexity of the Bell System; other U.S. firms which were highly centralized encountered diseconomies of scale when they attempted to run operations spread over the entire nation. See Alfred D. Chandler, Jr., Strategy and Structure: Chapters in the History of the Industrial Enterprise. Cambridge: MIT Press, 1962, pp. 225-282. But the Bell managers who discussed decentralization usually cited accommodation with the political system as their rationale. See Garnet 1985, op. cit., pp. 150-152.

20 Unless otherwise indicated, the statistics here and throughout are from U.S. Bureau of the Census, Historical Statistics of the United States, or Statistical Abstract of the United States. See also AT&T, Chief Statistician's Division, "Telephone Development in United States, 1876 to 1957", June 1958, in AT&T Archive.

21 Brock 1982, op. cit., pp. 153-155.

22 On systems engineering see Jay Liebowitz, Strengthening Systems Engineering: Kelly College at Bell Labs, Unpublished paper, Baltimore/Md.: Johns Hopkins University, 1983.

23 Peter Temin with Louis Galambos, The Fall of the Bell System: A Study in Prices and Politics. New York: Cambridge University Press, 1987, pp. 18-26.

24 McKinsey & Company, "A Study of Western Electric's Performance", March 1969; copy in AT&T Archive. Within the FCC in the late thirties there had been considerable concern about Western Electric's efficiency, and that position influenced the decision of the Department of Justice in 1949 to seek divestiture of AT&T's manufacturing subsidiary. But that issue had been settled, without divestiture, by the 1956 consent decree.

25 Dan Schiller, Telematics and Government. Norwood/N.J.: Ablex Publishing, 1982, pp. 8-15.

26 McKinsey & Company, "Meeting the Challenge of Competition for the Bell System Telecommunications Equipment Market", November 13, 1973; and "Interview," Donald E. Procknow, December 19, 1985; both in AT&T Archive.

27 Brock 1982, op. cit., pp. 198-286. Temin with Galambos 1987, op. cit., pp. 27-65.

28 John deButts threw down AT&T's gauntlet to the FCC and the Department of Justice (DOJ) in a speech, "An Unusual Obligation," delivered on September 20,

1973, at the annual convention of the National Association of Regulatory Commissioners. DOJ picked up the gauntlet in 1974.
29 Steve Coll, The Deal of the Century: The Breakup of AT&T. New York: Atheneum, 1986, Temin with Galambos 1987, op. cit., pp. 106-269. For a view from within (and near the top) of AT&T, see Alvin von Auw, Heritage & Destiny: Reflections on the Bell System in Transition. New York: Praeger, 1983.
30 Harry M. Shooshan III, (ed.), Disconnecting Bell; The Impact of the AT&T Divestiture. New York: Pergamon Press, 1984. W. Brooke Tunstall, Disconnecting Parties - Managing the Bell System Break-up: An Inside View. New York: McGraw-Hill, 1985. Richard H. K. Vietor and Davis Dyer, (eds.), Telecommunications in Transition: Managing Business and Regulatory Change. Boston: Harvard Business School, 1986.
31 Eli M. Noam, (ed.), Telecommunications Regulation: Today and Tomorrow. New York: Harcourt, Brace, Jovanovich, 1983. Leonard A. Schlesinger, Davis Dyer, Thomas N. Clough and Diane Landau, Chronicles of Corporate Change: Management Lessons from AT&T and Its Offspring. Lexington/Mass.: D.C. Heath, 1987.

CHAPTER 6
THE TELEPHONE IN FRANCE 1879-1979: NATIONAL
CHARACTERISTICS AND INTERNATIONAL INFLUENCES[1]

Catherine Bertho-Lavenir

1 Introduction

The development of the French telephone "system"[2] between 1879 and 1979 presented a kind of paradox: For although France was a powerful nation alongside the other economically developed countries, the state of its telephone technology was relatively inferior and the standard of equipment was a long way from the standard of such countries as the U.S. and Sweden and not even as high as the standard in Germany and Great Britain. This anomaly makes research into the historical development of the French telephone system interesting - especially when it is studied within a comparative perspective of the development of large technical systems. One should ask why this inferior standard of telecommunications should exist in France considering the fact that with respect to the general economic development France was equal to countries such as England and Germany. The reasons for this have to be found on the one hand within the sociological and political components of the development of these large telephone networks, and on the other hand in the way in which the national community allows for state intervention into the economic sphere.

A game with several actors

In order to have a full understanding of the problem, it is necessary to take into account not only the history of the enterprise or the administration charged with operating the telephone network, but also the "ensemble" of what can be termed the "System of Telecommunications". This system consists in the activities of essentially three kinds of actors: First the operator, then the manufacturers, and last the state, in its role as legislator. This point is all-important. One must bear in mind the separate identities of operator and administrator in France. Between 1879 and 1889 a certain number of urban networks, including the Paris network, which at that time represented the essential part of traffic and income, were owned by a concessionary company, the Société générale du Téléphone (SGT). Between 1870 and 1913 a significant number of submarine telegraphic cables were also entrusted to concessionary companies which after 1913 became state subsidiaries, although they continued to be governed by private law. In the same way between 1920 and 1953 the international radio telegraph communications between the most important nations (particularly between France and the United states) were secured by another concessionary company: Radio France.

The manufacturers who made up the second group of actors in this history of the French telephone are very different in terms of their economic and political standing. Competition existed only within the subgroups and in their field of competency. For example, the companies manufacturing switching equipment were not the same as those manufacturing transmission or telex equipment. Some companies had a greater influence than others. At this time, the most prestigious was the Swedish company LM Ericson. Since the end of the 19th century, this firm was the supplier of telephone equipment to the French public administration. Since 1911 it has owned a subsidiary in France, which, without playing the major role in the great industrial changes, still guarantees a certain amount of competition between suppliers of switching technology.

However, far more important, and symbolic, was the role of the American multi-national company ITT. Thanks to the matching of its products to the specific demands of the French, between 1923 and 1939 ITT almost achieved a monopoly as a supplier of switching technology. Its technical and political influence was enormous. In reaction to it, the government telecommunication engineers did everything they could to promote the emergence of a national industry. To do this they chose a "protégé": La Société industrielle des Téléphones (SIT), backed in the 1920s by the Banque de Paris et des Pays Bas. This company developed from a department of the SGT which had not been nationalized in 1889, namely the production factories. After taking over the electrical group CGE in 1927, the SIT failed to enter the switching market in the period between the two world wars and, consequently, the CGE head office lost interest in these activities for several years. After 1945 the situation changed. A type of alliance was set up between the governmental research center CNET and the SIT - which was now the CIT (representing the telecommunications department of the CGE). This alliance was later to produce spectacular results at the time of the development of the first electronic switching centers in the 1970s. CIT Alcatel secured a sufficiently strong position to allow CGE to purchase ITT's telecommunications subsidiaries around the world in 1985. In the development of the electronic switching equipment, however, CIT had met a French competitor, Thomson, in 1975. Thomson was heir to the group CSF which had been important since the First World War in another technical field: that of radio transmissions and later micro-wave links. In the transmission field, the government had since the 1920s at its disposal a national supplier SAT, beside the possibility of getting equipment manufactured on American patents. The French government deliberately encouraged the development of the SAT Company: Keen to escape Siemens's influence, the long-distance service decided in 1932 to order amplified telephone cables from SAT, a small company which had branched off from the Grammont firm. At that time these cables represented the most advanced technology. Greatly helped by the government, SAT had fairly rapidly acquired technical competence and industrial autonomy. After the Second World War SAT formed an alliance with SAGEM, which specialized in the construction of telex terminals. This created a manufacturing center which is still very significant today.

The third actor, the state, intervened in two ways: first as a regulator, secondly as a network operator. The legal basis of telecommunications remained remarkably stable throughout the entire period. In 1837 a law was passed which defined the status of the only telegraph then in use, the aerial telegraph. This law asserted the monopoly of the state over telegraphic transmissions and enabled the state to grant to private companies the rights to run all or part of the network. Step by step, the legislator inserted each new technological development in telecommunications as it came out into the 1837 law. In this way the telephone (in 1879), radiocommunications (in 1919), and later radio broadcasting and television entered successively the field of the PTT monopoly (although some were to leave later on).

In practice regulatory power is distributed at several levels within the state: first there is the government level, then the ministry and, finally, the "administration" of the PTT. Laws and decrees emanate from the government. The ministry, which passes decrees, ministerial orders and makes decisions, can give some companies the right to become operators. But it is the administration proper which regulates things on the borderline between state action and private enterprise, such as the installation and maintenance of automated switchboards on customer premises. The "administration" (i.e. "department", but distinctly separate from the ministry) thus found itself, before the recent evolution, both in the position of judge and of participant.

Of course, this distribution of powers is broadly the same as in other European countries. It does not in itself account for the long crisis of the telephone in France, which has more complex roots.

A perpetually refueled crisis

The history of the French telephone system, in fact, is dominated by a "crisis" situation. The stages of its evolution can be divided into two periods of unequal significance. The first period runs from 1879 to 1975. Contemporary opinion perceives this as being almost a century of continuous crisis. However, one can find three different kinds of reasons for this crisis, connected to different stages in history. The years before the 1914 war (1879-1914) were dominated by problems of institutional order. The telephone development was impaired by the status which it was given: first the legal status and, secondly, the financial one. Both were unsuitable. A second stage which runs from 1919 to 1939 was dominated by manufacturing problems. A third and last stage between 1945 and 1975 is characterized by the split between the progress in research connected to the undeniable take-off of the national industry on the one hand, and the persistent state of crisis for the telephone users on the other. 1975 marks the beginning of a new era for the telephone. It is signalled by the fact that by that time, France had caught up with other industrialized countries in terms of telecommunications public equipment. The effort made to solve this long-lasting crisis had been fruitful, but from 1985 on, new problems arose.

The analysis of each of these stages raises three questions, which are the same for each period: What is the relationship between the actors "within" the state? What is the relationship between the state and the industry? And, finally, how does the telephone rate in public opinion? The answers reveal two apparently contradictory phenomena: On one side the crises of the French telephone are directly related to political or sociological trends specific to France, but at the same time the great stages of the general system evolution are similar to those in all European countries. For example, the telephone in France, as in Germany and Great Britain, had to develop in a context characterized by the power of the telegraph administration already in existence. Similarly after 1919, the problems of telephone development acquired an essentially industrial dimension and investment capacity became a crucial factor in all European countries, simply because they found themselves facing identical technological innovations which forced them to update their central stations in the big towns and also their long-distance communications. One could therefore formulate the following hypothesis: Within the history of these large technological systems there is some kind of dialectical relation between the factors related to science or international economy, which were therefore common to all countries, and consequently posed the same problems and brought about the same solutions, and the factors particular to each country which depend on its economic character, its political life and even its cultural traditions. From the balance between these different factors there emerges in each European country a particular national profile of the telecommunications system. This gives rise to systems which are both totally original and yet fairly similar to each other.

2 The interminable crisis of the French telephone (1879-1950)

A legal and financial crisis (1879-1914)

In France, as in England or Germany, the heritage of the telegraph greatly affected the way in which the telephone was introduced, both in the technological approach chosen and in the institutional mechanism adopted. In 1878 the French state had merged the telegraph administration and the postal administration. There were some economic reasons for this decision: in order to attract new users, the telegraph had to penetrate the rural areas, and it seemed sensible to use the post office networks in order to do this. However, there were also political reasons involved: the recently installed Republic was quite keen to place the reputedly Bonapartist telegraph administration under the authority of the postal administration, which was larger in numbers and more loyal to the Republican cause. This led to a phenomenon specific to the French case: a lack of management personnel in the telegraph administration. A great number of engineers left their jobs because they were given less responsibility just at the time the administration needed better staff able to deal with the new invention of the telephone, for which the patents had been registered in the United States in 1876, and in France and in Europe in 1877.

The unsuitability of the regulatory regime

In February 1879 the Minister of the new PTT Ministry made his decision public: The construction of the telephone system was not to be undertaken by the Administration, as the telegraph system had been, but was to be granted, in each city, to the companies which asked for it. There are several reasons for this. The first is technical: long-distance telephone transmissions were not yet feasible at that time, quite in contrast to the telegraph, where problems of fading did not occur. The telephone network was not perceived by the authorities as a strategic network covering the whole territory, and the political or military necessity to entrust the development of the networks to the state was therefore not even considered. In contrast, the last quarter of the nineteenth century saw the development of many urban networks: networks for water distribution, electricity, and tramways, which for the most part were granted by towns to specialized companies. It seemed logical and simple to proceed in the same way with the telephone. And so, in 1879, the task of constructing the Paris network was granted to the SGT, which also obtained the concession for Rouen, Le Havre etc. For ten years, between 1879 and 1889, the main French telephone networks were private.

In 1889 this liberal period ended in a brutal nationalization, the administration of the PTT taking over all networks. What had happened? First, a phenomenon of the economic logic of the networks: following very rapid mergers, there had been a movement towards a monopoly situation for the private concessionary company, which was frowned upon by public opinion. Other factors must be taken into account which are specific to the French political culture. On the one hand, suspicion prevailed as regards the motives and ability of any concessionary, and it was widely assumed that they would unduly profit from their position. On the other hand, ministers and politicians did not want to be reproached for having diminished the scope of public control. The SGT's concessions were short (4 years) and restrictive, which all pointed to the possibility of being taken over by the state. Just at the time when the company tried to change the concession to a longer lease (25 years), which would at least have allowed it to invest, the Minister of Finance who had signed the lease was ousted. His successor and the government voided the settlement. On the whole, the Administration des télégraphes did not loyally play the game of concessions but started to compete with the concessionary company by building its own networks in certain towns.

Under these conditions, the SGT quickly stopped investing; the rates were raised and performance standards dropped. At the time when nationalization was being talked about, few people actually spoke out, even in the business world, to defend the SGT. Nationalization itself was not perceived as being ideological, and it took place amidst relative indifference. One can think this means that telecommunications was considered to be fairly neutral by ideologically-minded parties. But it can also be said that a kind of secret consensus was forming: It also concerned the business world, which favored handing over to the state the responsibility of insuring the functioning of the network. This was thought more likely to guarantee equal access to all consumers than a private company - this being necessary for the maintenance of fair competition in the economic world.

In fact, the nationalization of the telephone was never seriously contested, even in the 1920s when the Minister found himself confronted by a takeover bid for both the equipment and the operation of the French telephone from two competitive groups. The first came from the Banque de Paris et des Pays Bas, which was connected with the SIT; the second came from the newly created American multinational company ITT. The Minister replied to Colonel Behn, head of ITT, saying that he could not accept because, "It would be contrary to the Republican traditions of the country." In plain language that meant that the Minister did not wish to confront the influential and attentive PTT workers' unions over this issue. But at the same time he was conscious of the fact that within the country there did not exist any current of opinion likely to support such an action. France in this respect was by no means in a unique position in Europe. All the telephone and telegraph networks, except those in Denmark, were run by the Administrations. Even Victorian England, homeland of liberal theory, went through three nationalizations: the telegraph in 1871, intercity lines in 1896, and urban telephone networks in 1912.

The financial crisis (1889-1914)

This "secret" consensus, however, was not strong enough to support a dynamic state policy, as had been the case at the time of the construction of the telegraph network under the Second Empire. The second aspect of the telephone crisis between 1889 and 1914 regards the contradictions in its financing. The Members of Parliament had put the construction of the telephone network back into the hands of the Administration, but they refused to appropriate at the national level the funds needed for these investments. They expected the local communities to provide the funds: Towns wanting a telephone network would advance the necessary credit to the administration. In practice it would often be small banks[3] who would make the funds available upon request of the Chambers of Commerce, which were acting as intermediaries. The legislator[4] believed that this practice would balance supply and demand from the start, but in fact it was to have perverse effects. Besides, this practice was not conducive to the real extension of the network. What interested the notables in charge of allocating public funds at the regional level was not the network for local communications (as one can see from reports on the debates held in regional assemblies), but the "wire to Paris". This lead to a Malthusian attitude once the first subscribers were served: The extension of the network was rejected so as not to cumber the precious wire to Paris. This financing method was also unsuitable for the construction of interurban communications. It was here that important developments became possible, when induction coils, known as "Pupin coils", permitted sound amplification. Finally, each local community tended to ask for "its" own network. This is why the French network is made up of a large number of tiny networks (14,000 as opposed to only 7,000 in Germany). This made long-distance communications impractical. In order to assure the financing of the long-distance communications, which the local communities were not much motivated to do, the engineers started a merger of the receipts of different networks, which had the effect of inhibiting any profitability measurement of a network.

One could ask, why was the telephone not offered the massive and centralized investments which could have been expected? It seems that the answer has to be found partly in facts belonging to the sphere of collective psychology, or culture, and partly in the existence of a working telegraphic and postal network. There lies an important cultural fact: In the France of the *Belle Epoque*, the telephone had a rather licentious image for its non-professional uses[5]. "Why the telephone?" the writer Colette is supposed to have said, "It is only used by men for serious business or by women who have got something to hide." In this last instance the state could not be the financial backer of an enterprise which was not seen as legitimate in terms of the morals of the time. More profoundly, it seems that the French society did not really need the telephone to complete its communications system. In this respect there was a big difference between France and the United States, where Vail could make universal service a mobilizing theme because of the vast spaces in America yet to be conquered, and the large number of completely isolated rural households. The problem was far less vital in Old Europe, at least in France, where there had been a daily domestic mail service to every community since 1832[6], and where the telegraph network meant that one could contact someone anywhere in the country in a few hours, and at very little cost. All these reasons combine to explain the authority's apathy as regards the telephone. In 1906 and again in 1909 a social and a technical crisis occurred together. Big strikes, the first civil servant strikes, paralyzed the network. In 1910 big floods in Paris showed up the technical[7] inefficiency of those responsible for the network. Several studies or reports clearly identified the roots of the problem. However, no decision was made about this sector as it was generally considered as being non-strategic: The telephone as an instrument of civil society was not given priority in those years when preparations were being made for war.

The manufacturing problem (1919-1939)

After the war, the telephone problem took a different direction: The time had come to install automated switching centers in the big cities, such as Paris, Berlin and London. Moreover, the amplifying valves (produced mainly by American and German manufacturers) permitted planning the creation of interurban telephonic cable networks. All this required considerable investments. It was difficult to obtain adequate funds for an administration whose inefficiencies had been demonstrated in the pre-war crises. Yet there existed a general consensus concerning the need to modernize the institutional context of the telephone. What happened was adaptation and not radical change. In 1922, when Mr. Behn proposed to the Minister of the PTT to take over both the supply of telephonic equipment and the network itself (a proposition he had already made or would make to Spain and Belgium), the Minister of the PTT, as noted already, refused as he had refused a similar offer from the SIT. The reform went two ways: one was the return to concessions and the other the modernization of management methods. Recourse to concessions had been largely practiced in the last quarter of the 19th century for the development of telegraphic submarine cables. However, bearing in mind a series of bankruptcies of French corporations, the Administration began to buy back the cable companies which had been functioning as subsidiaries to the state since 1913, i.e. as private companies with public funding. The Administration's intention was to set up lines which were regarded as being strategically important (especially those to the colonies) and which were in danger of being taken over by English companies.

By no means discouraged, the Minister of the PTT decided in 1920 to use concessions to develop a new technology, radiotelegraphy, largely dating from the war, which had opened a promising market, particularly the transatlantic link. An agreement was made between the Minister L. Deschamps and E. Girardeau, head of the CSF, a company manufacturing radio equipment which, during the war, had achieved a technical ability which gave it worldwide fame. Concluded at a time when there was a right-wing majority in the National Assembly, the Deschamps-Girardeau agreement caused a scandal. It met with hostility from the PTT unions and the services themselves who felt deprived of a certain number of links they would have liked to supply for themselves. In 1924 the return of the left-wing coalition to Government allowed the agreement's opponents to considerably limit its application, without going as far as annulling it. The result was, that at the time when international competition was fierce in this field, France had one company, Radio France, which was not very dynamic, and which was being permanently restricted by the concessionary administration.

At the same time, France proceeded towards a modernization of the PTT management. Since nationalization in 1889, the telephone budget had been administered like any other department, according to laws passed under Napoleon I, which particularly specified that expenditure could not be spread over several years. In the same way it was forbidden to set aside an income for a particular expenditure. In practice, the postal revenues, like those of the telephone, disappeared into the vast mass of taxes, and the financing of the Post Offices and the telephone centers had to compete with the expenditures of the Army, the Department of Justice and Public Education. Some new provisions were adopted at the time the budget was voted in 1923. These aimed to enable the PTT to move towards a more "industrial and commercial" management[8]. The PTT's budget was separated from the general state budget, authorization was given for borrowing, a sinking fund had to be created, all this in view of the investments necessary in order to install automated telephone exchanges in the large towns and to build a network of amplified cables. But what happened to the Deschamps-Girardeau agreement also affected the 1923 budget. After 1924, under pressure from the Minister of Finance in particular, the reform was distorted, although not completely annulled. The budget annex remained, but the promised sinking fund was never created. The multi-annual investment plans were compromised by the budgetary deficit; moreover, the merging of accounts with those of the Post Office meant that economic calculations and the forecasting of investment profits could not be made.

Under these conditions, the Administration des télégraphes et téléphones was not in a strong enough position to resist the pressure from foreign manufacturers. Thus, two relational patterns between the state and the manufacturers developed, differing according to which technical sector they applied to: In switching technology ITT had complete technical dominance until the Second World War, whereas in transmissions an alliance was established between the Administration and a national manufacturer in order to acquire a certain amount of technical independence.

The way in which ITT monopolized the French switching technology market between the two wars revealed the importance of the cultural dimension in decisions which appeared to be purely technical. ITT bought factories in France, utilized and turned into cash the patents of famous engineers employed in these firms, and integrated the specific demands of the Administration into its proposals, for instance the preservation of the old and familiar names of manual switching stations upon automatization. ITT also set up a research laboratory in Paris, which was situated right in front of the Ministry in order to demonstrate that research was being carried out in France[9]. Lastly it conducted a determined lobby action directed at the technicians in the head office of the telephone services. The result was striking: ITT gained the commission for two kinds of switching systems, the Rotary for the cities, and the R6 for the average-sized towns implemented in 1926 and 1929. The only condition was that ITT should authorize other companies with factories in France to manufacture under license.

In the transmissions sector, the situation was quite different. Right from the start the military authorities considered the amplified telephonic cables to have strategic value. Again we see the difference between an urban network and a national network, which was already evident when the telephone was introduced. The existing equipment was manufactured under either American or German patent. One of the first French amplified cables, Paris-Bordeaux, was delivered and installed by Siemens without the technical intervention of the French Administration. Further German penetration was, however, held up somewhat because the French manufacturers were hostile to the practice of reparations (in the case of the telephone switching system there were speeches from the *députés* in the Chamber, asking that the systems manufactured under German patent should not be retained). They feared indeed an invasion of their market. The role of government engineers was also very important. They - and this tends to be a general characteristic of large technical systems - were naturally in favor of all the projects which would grant them a highly respected professional status, whether they were working in industry or for the Administration. In this particular case their interests were similar to those of the military. In 1924, the telephone and telegraph Administration set up a special service for the installation of amplified cables, the service for long-distance lines, which partly escaped from heavy bureaucracy - not only because of its technical character, but also because its projects were by nature outside the functional and hierarchical structure of the PTT, a structure which reflects the regional division into *departements*, where PTT authority lies in the hands of the regional director of postal services. This new service was specialized in electronics, so it attracted the brilliant young engineers. An alliance was then established between the LGD service and a small French company partly set up with the help of the Administration, a company made up of a small team of highly specialized engineers from the Grammont factories. The Administration ordered cables which were of entirely French technology. This demanded a close collaboration between the operator, i.e. the Administration, and the manufacturer to ensure the best results. Even to the contemporary witnesses the border between "public" and "private" became totally blurred, both in the everyday life of the engineers and in the economic behavior of the actors[10]. For example, during the first years, the Administration accepted deliveries of mediocre quality cables which they would not have accepted from other suppliers, in order to support its partners' apprenticeship. Between the two wars a technical system thus evolved which so closely intertwined the operating Administration and the national industry that it was difficult to know where the decisions were made, and who had the power to make them.

Despite the ambitions and the technical ability of a few engineers, the French telephone network remained mediocre: The number of subscribers, and above all the quality of service remained low. As we have seen, the attempts to reform the barely established institutions were either delayed or rejected. The reason is probably due to public opinion which put no real pressure on its representatives. The telephone did not spread beyond the wealthy areas, and the literature available to the general public tended to provide the reader with visions of charming young girls working in the telephone exchange while automatization was the technical challenge of the time. This new development was also met with reticence by users who were accustomed to being waited on and who identified manual operation with domestic service. This went on with no obvious change, and even as late as 1935 it was as if French society had become accustomed to the inadequacies of the telephone system.

3 The engineers' telephone (1945-1975)

To a certain degree, the period following the war was similar to that just before its outbreak. A total divergence was evident in France between the user's position, which remained unchanged until 1975, and that of research and industry which became more and more advanced. The telephone users were faced with a shortage of lines, which they accepted fairly well for a long time, while at the same time public research and national industries were joining up in order to elevate French technology in other sectors to an international standing.

An accepted shortage? In 1967 the standard of the telephone in France - compared to other countries - was no better than it had been before the war. With 3.9 million principal user lines and 8 million extensions of all types, the French network was no bigger than those of New York or Chicago. France was 28th in the world for the rate of automatization, and 16th after Australia, Finland and Japan for telephone density (number of extensions of any type per capita).

This situation provoked sneers from the satirists: It no longer appealed to the mentality of the young. If one carefully observes the way in which the telephone is treated in the "new wave" films of the time, many scenes will be noted where teenagers have to get out to make a telephone call, away from their parents' ears, on the landing, on the balcony, or at the neighbors' house. However, no movement capable of alerting the political authorities was apparent. In 1964 an opinion poll[11] asked the French what they thought the most important thing to possess was: The telephone came in sixth and practically last, just before the record player, but far behind the car and domestic electrical appliances. A Minister from the PTT defended his modest budget in front of his own majority, assuring everyone that the telephone was a gadget that the French did not really need. It was hardly surprising therefore that the telephone was not given priority in any of the post-war plans.

For this reason, telecommunications research and development were not really led by the market. After the war, industries all over the world suddenly had to revert to civil production. Research into high frequency resulted in the development of micro-wave equipment and television facilities. From the moment television sets existed, the needs for transmission capacity rose. In 1962 the first civilian telecommunications satellite transmitted images across the Atlantic, at the same time as telephone conversations.

Inspired in a way by this technical competition, the manufacturers of submarine cables started to install the first transatlantic telephone cables in 1956. At the same time, two generations of switching technology began to replace the pre-war electro-mechanical equipment. From the 1950s, the more rapid crossbar systems substituted the pre-war rotary systems. At the same time, the United States, France and Great Britain were struggling with the production of automatic electronic switches. After tests it was shown that the central exchanges required reliable and high-performance data processing. The Americans and the British, who already had a headstart, then met with endless difficulties in the development of their systems. The French had been left behind, but they happened to take off at the right moment considering the performing capacity of the components which were then on the market; since 1975 they have been able to present two reliable systems.

In fact, although the French Administration had so many difficulties with respect to its subscribers, it paradoxically was involved in the great post-war technological events. In 1962 Bell Laboratories launched the first civilian telecommunications satellite. They needed to find partners of the same technical skill on the other side of the Atlantic. England was the traditional partner, and France, when it asked AT&T to set up a satellite station in Brittany, became the second one. After 1971, within the domain of submarine cable, the French Administration and the manufacturers, until then just partners, entered at the research level into consortiums which were involved in the construction and operation of transatlantic cables. Finally, in the terminal equipment field[12] there emerged a French telex company, which later dealt with electronic telexes, which constituted the major part of its export figures at a time when all the French teleprinters were still of pre-war English origin (Creed).

At the beginning of this slow process on the way to technical autonomy, there was a close cooperation between the Administration and the national industry which has to be carefully checked to be in accordance with the rules of the market. There is something in this that speaks of the typically French tradition of "grands projets". This practice links a team of State engineers, who are completely free to carry out a project defined by technical requirements, with a national industry, which receives orders in which the price is not the only criterium.

In fact, the technical competence and autonomy of the Administration was increased by its role as buyer vis-à-vis industry. Between 1940 and 1944, the Vichy Regime had clearly contributed to telecommunication's emancipation from the supervision of the Post Office. The telegraph and telephone branches were reorganized into one telecommunications administration in 1941, which became the Direction générale des télécommunications (DGT) just after the war. Also in 1941 the research services, which had been dispersed until then, were reassembled into one center. After the war two structures were set up to regulate in a rather dirigistic fashion the economic relations between the Administration and the industry. In 1947 there was a regrouping, within a company of the mixed economy type, of the transmission manufacturers and the Administration, and in 1958 the other reassembled the switching technology manufacturers, but with less success.

This rise in power of the technical and manufacturing sectors had no equivalent as regards the subscribers. After 1945, the priorities in investment and equipment matters were defined by the planning authority. The telecommunications engineers did not succeed in introducing the telephone as a priority into any of the first four plans. The blame lay in the weakness of their institutional status: within the functional hierarchy of the Ministry, within the regions and departments, it was always the Post Office men who had the experience and the responsibility for relations with the elected representatives. In the 1950s, the political weight of the telecommunications engineers was practically non-existent. They were not represented in ministerial cabinets outside their own ministry, and their ability to intervene at the important arbitrations was very restricted.

This situation was to change in two steps: in 1967-1968, and then again in 1975. Under the presidency of Georges Pompidou, some decisions affecting the institutional order were taken which allowed the restrictions set up at the end of the 19th century to be withdrawn. First French Telecommunications were authorized to borrow in the foreign market (1967), then specialized finance societies were set up (the first was introduced to the Stock Exchange in 1975); the management of the Telecommunications Service was considerably modernized, on the model of company management; and the accounts were separated from those of the Post Office.

What were the reasons that contributed to this sudden awareness of the political world? Partly the pressure of public opinion, reflected by their representatives. Without any outside stimulation, from 1961 to 1965 the number of non-satisfied demands for telephone lines rose from 119,000 to 407,000. In 1969, articles appeared in the newspaper Le Monde, condemning the telephone crisis; a group of ambitious young engineers informed the political world, and proposed some solutions. By and large the industrial make-up of the country was transformed - and lastly one should not ignore the politics of industrial groups. The CGE in particular greatly participated in the post-war installation of high-voltage electricity, and it certainly did not want to do away with a new range of public markets in telecommunications.

Table 1: Telephone Main Stations:
Number in Thousands and Density per 100 Inhabitants

Year	France		Great Britain		F.R.G.	
1938	1,001	2.4	1,857	3.9	--	--
1950	1,442	3.4	3,043	6.0	1,419	2.9
1960	2,194	4.8	4,647	8.9	3,221	5.8
1970	4,144	8.1	8,380	15.0	8,700	14.1
1980	15,898	28.9	17,696	31.7	20,535	33.4
1985	23,031	40.8	20,921	38.3	25,588	41.9

Sources: M. Correze, "Rapport au nom de la commission de contrôle de la gestion du service public du téléphone", Journal officiel (1974); Annuaire statistique des télécommunications du secteur public (14e édition). Geneva: Union Internationale des Télécommunications, 1987.

Between 1969 and 1975, an improvement in the network took place. After 1975 this gave rise to a huge growth in the number of subscribers. There were 6.2 million subscribers in 1974 and 20 million in 1982. Twice as many lines were installed in eight years as had been installed in the preceding one hundred years. This did not occur without producing tensions. Within the PTT the rearrangements planned in 1968-1969 proved to be sufficient; it was possible to direct the growth without having to change the status of telecommunications. A proposed law drafted in 1969 by Valéry Giscard d'Estaing, creating a national telephone company, was not approved. In contrast, tensions were acute within the industrial system. In 1974, with V. Giscard d'Estaing as President, what had become a traditional alliance between the Administration of Telecommunications and its research capacity on the one hand, and national industry on the other, was to be deeply reconsidered. A pattern which had worked smoothly through the Fourth Republic governments and those of General de Gaulle and Pompidou was deemed obsolete. Looking towards the international markets and being convinced that the era of the construction of a national industry had come to an end, and that it was time to give this industry an international perspective, the new government reduced the role of the public research center after 1974. By doing this, the government cut itself off from a whole generation of engineers who had the same fears as the PTT personnel. In 1974, a very long strike (much more serious than that of 1968) shook the PTT. For the first time, the researchers at the Centre national d'étude des telecommunications joined a movement started by the Post Office workers. And yet, the position of the unions was not simple for Telecommunications, partly because since the 1914 war, they had made a dogma out of the unity of the Post Office and Télécommunications which offered strong advantages and guarantees to the small employee. Furthermore, after 1974, the construction of the telephone network had benefited from unprecedented investments which were to guarantee an unexpected expansion of the public service, since the formidable growth had not been accompanied by a deep change in status.

At this moment, a kind of unacknowledged alliance was established between the personnel and the management of telecommunications, centered around a common objective: to carry out a large technological project, the eventual achievement of a nationwide telephone service. This would be the realization of an ideal common to all those working in the technical system of the telephone.

In fact, since 1975 those in charge of the system had had powerful means available. Their technical project was neither contested nor controlled (who else could judge the validity of the choice of switching technology?). The only oppositions were internal. To their benefit, there was also a consensus of the national community concerning the beneficial nature of their enterprise. All the telecommunication agents who participated in these years of the "great leap" look back upon them nostalgically as having been a golden age when the telephone Administration finally achieved its vocation - a situation which recalls that of Electricité de France in the 1960s.

However, this period of accomplishment brought with it the seeds of further problems. It was quickly realized that the moment would come when the whole country would be equipped with telephones. On the other hand, the prevailing notion of a necessary union between data-processing and Telecommunications forced the telephone personnel into taking interest in new forms of activity. The DGT had been operating telecomputer services since the 1960s, either directly or via subsidiaries. In 1978 it launched a programme of equipping France with Videotex. After a few ups and downs, this proved to be a success. However, its painful birth made the characteristics and limits of the telephone system clear. With its renewed institutions, mobilized by a cause highly congruent with the professional culture of its members, this system had been perfectly adapted to a great project such as equipping the whole of France with telephones. The completion of the Videotex project, in contrast, called for new skills: an aptitude for social bargaining, the ability to conciliate the interests of the press, the representatives, and the manufacturers, and to be aware of market fluctuations and the variation in tariffs. Telecommunications staff were rather new in these fields, and yet they managed, at the cost of deep changes in the existing value system. And this in France in the 1980s, at the very time when deregulation was the word, reopening conflicts similar to the those that have been described.

4 Conclusion

Are there conclusions to be drawn from this unusual history? It seems that the history of the telephone in France provides a clear indication as to the general aspects and the more particular qualities of this technical system. First of all, the telephone is clearly not a product like any other. On the one side, as a communication network it is interesting to states for military and security reasons, on the other side the networks, through high concentration and mergers, tend to bring out monopolies. The European states do not tend to tolerate private monopolies. For this reason the telephone system operates in the form of a public monopoly managed by an administration connected to the postal service in most European countries. In contrast, national specificities appear in the chronology and the way in which these administrations were set up. Prussia directly entrusted the running of the telephone system to the postal and telegraphic administration; France did not make up its mind until ten years of concessions had expired. England, even less of an exponent of planned economy, instituted two successive nationalizations. In the United States, since the 1920s, AT&T had to affirm, through a policy of all-embracing communication, that it would carry out its public service obligations in order that its size and monopoly would be tolerated by society.

Later on the successive adaptations were always made in response to the blow of technical innovations whose chronology was the same for the entire industrial world. In 1920 it was necessary for Europe to be equipped with new centers and long-distance telephone lines. At the same time, each country was moving towards a reorganization of the management of its telephone administration: tariff and budget reform in Germany, budgetary reorganization in England and France, Italy's adoption of a decentralized organization, taking advantage of the fascist economy's own institutions, etc.. In this way the European countries responded with their own institutional instruments suited to their own interests and political traditions. These were particularly important in a sector where the idea of national interest in the form of an industry was combined, in 1914 and particularly in 1945, with the perception of strategic interests. This allowed for the development of an industry for a single buyer, fairly free from the influence of the market forces. This was not unique or specific to France. In the United States, AT&T as the sole operator was supplied by one company, Western Electric, which it owned, together with the research laboratories - Bell Labs. In England, Germany and Italy, the administrations prefer to place their orders with the national industries. As a consequence, every country facing industrial and technological changes must at the same time face the necessity of adapting its organization as well. Each one responds within its own tradition. England, impregnated with classical liberalism, was the first to restore "enterprise" status to its operator. Germany had decided until recently to preserve the original status of the telephone network and the basic structure of the system. France proceeded by rearranging the periphery of the system, granting concessions to private companies, for instance for the radio-telephone or networks for value-added services, but putting them into competition with the administration's services. Finally, the particular cultures of the large technical systems should be taken into account: What will the determining factor in the amendments to the present borders between data-processing and telecommunications be - the solidarity of national interest, threatened by the internationalism of the large firms, or the complicity of the telephone workers when faced with the data-processors (should these groups remain distinct)?

What is the most important factor: the international evolution, or national characteristics? This is difficult to answer. In France for example, the statistical trend of the country's telephone density seems to faithfully reflect a specific national history marked by particular political conflicts: between the union and the Post Office in 1878, initiated by the Third Republic; liberal pressure in 1922 which was thwarted by the left wing coalition in 1924; the modernization of France by Georges Pompidou between 1968-1975. But this notion of specificity allows at least for two comments. First, if one takes a long-term point of view on things, one can see that the rise and the definite establishment of the telephone system as a whole had about the same magnitude in France, Germany and England[13]. At times and in some respects, France could be termed "backwards", but that did not impair the development process in the end. Within the industrialized countries, the dividing line lay between the US at the head, and the Scandinavian countries and Switzerland on one side and Germany, England, Italy and France on the other.

A reading of the long-term telephone statistics leads to the same observation; France's evolution, as regards telephony, was not aberrant when compared with other nations (except perhaps in the post-war years): "At the end of the war we tried," wrote Jean Voge, chief engineer at Télécommunications, "to base our predictions for the growth of the telephone network on extrapolations of the past. Between 1889 and 1933, we saw the number of subscribers rise from 11,000 to 860,000. The annual growth, which was progressively reduced from 15% to 7.5% (apart from the war period 1914-1918, when there was total stagnation), was at that time quite similar to that of other European countries and that of the United States. We could therefore expect this growth to follow a pattern of about 7% a year, corresponding to a doubling in the number of users every ten years. Therefore from 860,000 lines in 1933, we could expect to reach 24 million after fifty years. And in fact it will more or less be so, as this goal will be achieved by the end of 1986."

This may allow a conclusion: Although the details concerning the organization and development of the telephone system were so blatantly distinct from one country to another, it looks as though the same main current carried along the countries considered.

Notes

1 Translated from the French by Ginny Lacey.
2 Catherine Bertho, Télégraphes et Téléphones. Paris: Hachette 1981. See also Louis J. Libois, Genèse et croissance de télécommunications. Paris: Masson, 1983.
3 Jeanne Leroy-Fournier, "Le téléphone a cent ans ... et 80 ans à Poitiers", Bulletin des Télécommunications Poitou-Charentes, (1976)5, pp. 13-16.
4 Jacques Attali and Yves Stourdze, "The slow death of monologue in French society" in I. de Sola Pool (ed.), The Social Impact of the Telephone. Cambridge/Mass.: MIT Press, 1977.
5 Yves Stourze, "Généalogie des Télécommunications françaises", in A. Giraud, J.C. Missika, and D. Wolton (eds.), Les réseaux pensants, télécommunications et société. Paris: Masson, 1978.
6 Catherine Bertho, "Les réseaux de télécommunication au XIXème siècle", in La communication dans l'histoire, Tricentenaire de Colbert. Colloque de Reims 1983, Travaux de l'Académie nationale de Reims, Vol. 164, Reims, 1985, pp. 155-166.
7 Catherine Bertho, "Histoire des réseaux téléphoniques parisiens, 1879-1927", Annales de la recherche urbaine, (1984)23-24, pp. 143-155.
8 Paul Laffont, "La crise du téléphone", L'Information économique et financière (1922) n° spécial. "Projet de loi relatif à l'amélioration et à l'extension du réseau et de l'outillage téléphonique à Paris et dans les départements", exposé des motifs par Yves Le Trocquer, ministre des Finances. Journal officiel, séance du 22 décembre 1921.
9 Maurice Deloraine, Des ondes et des hommes. Paris: Flammarion, 1974.
10 Revue Câbles et Transmission (1948)1. See also Georges Fuchs, "La SAT avant 1940, la collaboration exploitant - industrie dans le développement des câbles amplifiés", paper presented at the conference "Droit et histoire des Télécommunications", November 1987, to appear in the "Collection de l'Ecole pratique des Hautes Etudes en Sciences Sociales", DROZ, Genève.
11 Sondages "Les PTT: Le public et le personnel", Revue française de l'opinion publique (1965)2/3, p. 77.
12 Philippe Leger, "Le télex, naissance et croissance de l'activité télécommunications nationale et internationale dans une entreprise française", paper presented at the conference "Droit et histoire des télécommunications", 1987 (see Note 10).

13 See Libois 1983, op. cit., p. 83: "Situation du téléphone dans quelques pays à la fin de 1938", densité (postes de toute nature)

Etats Unis	15.27
Suède	12.67
Grande Bretagne	6.74
Allemagne	5.20
France	3.79
Italie	1.43

CHAPTER 7
THE POLITICS OF GROWTH:
THE GERMAN TELEPHONE SYSTEM

Frank Thomas

1 The telephone as a large technical system

The aim of this paper is to portray the development of the German telephone system as it resulted from the interaction among a set of corporate actors. The course of the development cannot be inferred from the impact of a single variable. On the contrary, system development is seen as the result of decisions (as well as of non-decisions), choices among a number of alternatives made by a definable set of actors. The choices made are constrained by the actual environment of the actors and by the sediment of previous decisions, but they are not determined by any of them.

The telephone system is a technical system because its central function is the transmission of spoken information by electrical waves. It is a large system because of its sheer size in manpower and capital, and because in an advanced state of development it encompasses most of the territory of a society. Finally, the different components that make the telephone work form a system - they are all needed and they interact.

In reconstructing the development of the German telephone system, I shall highlight structural and environmental aspects as well as the temporal and spatial dimensions of the process:

- the embedding of the telephone system in the overall communications and transport system;
- the special weight of political actors and of their political and economic decisions;
- the time-consuming, stepwise integration of separate elements into a single system;
- the importance of an analysis of geographical properties of a

technical system, especially if the technical system has a communications function.

For the purpose of intellectual parsimony, five central phases were selected to represent the development of the German telephone system:
- The introduction of the telephone shows the starting conditions which greatly influenced the further development.
- After a decade of growth, the first difficulties arose that were tackled by two laws and a parallel change of technology.
- During the Weimar Republic, the first steps were taken to give the postal and telecommunications system a certain degree of autonomy.
- After 1933, a reshaping of the actor network and a change of function resulted in a massive geographical spread of the system.
- The reconstruction and expansion after the Second World War implies a change of functions that enormously accelerated the growth of the system.

As space is limited, all references to the interaction with the economic environment, with international actors, and all but the most superficial remarks about the technical development of the system are omitted. Nevertheless, this does not mean that they are unimportant.

2 The introduction of the telephone

The telephone entered into the German postal communications system in a different way than in most other countries. It entered in a two-stage process: In 1877, it was first used as an auxiliary telegraph apparatus. The idea cherished by the German Postmaster General, Heinrich Stephan, to open subscriber telephone networks was not carried out because of a lack of demand by customers. Two years later, the second stage began when private businessmen took the initiative and asked the German administration to get concessions for private telephone networks. Only then the administration felt moved to act. It declared private telephone networks to be unconstitutional and opened state telephone networks with public access.

The setting for the introduction of the telephone in Germany was an economy in a state of full industrialization. Growing societal and spatial differentiation led to an expanding traffic and communications sector. Urbanization and the clustering of enterprises in the cities made it possible to use the early telephone with its limited transmission range[1]. But most city centers were still small in area, so that personal visits, urban mail, and messenger services were able to compete successfully with the telephone. In the political arena, Germany had been united as a constitutional monarchy only since 1871. Its central government remained weak in comparison to the state governments. For instance, separate postal and telegraph administrations existed in Bavaria and in Württemberg. The finances of the Reich were also weak so that the income derived from the Imperial Posts and Telegraphs (RPTV) was important for the central government. Both agencies were merged only in 1875. They were controlled by the Reich Post Office, the precursor of the later Post Ministry. So the field of communications has always been dominated by political-administrative actors. When later on the telephone was declared to be a part of the telegraph system, no special organization needed to be set up for it.

In the second part of the 1870s, the Post Office modernized and geographically extended its communications networks into rural areas. Its intention was to improve political and economic integration, to strengthen administrative control of the territory and its population and to put an end to the financial losses of the telegraph system. An important obstacle in implementing this policy were the marginal returns from telegram use that were expected in peripheral areas[2].

In Bavaria and in Württemberg, where state railways and the separate postal and telegraph services were administered within the framework of the same ministry, apparently neither a modernization nor an extension of the state communications system was perceived to be as necessary as the Reich Post Office held it to be. Here, industrialization was still at a low level, and the economies of scope that were inherent in this integrated type of state bureaucracy kept the running costs of telegraph stations low. Where the RPTV was interested in network expansion, the southern German telegraph administrations therefore were not. This divergence of goals partly explains the different time of adoption of the telephone in the three areas.

The first stage of telephone introduction began when Heinrich Stephan, head of the RPTV, heard about the invention of a "speaking telegraph" in October, 1876. At first, he did not react to the news because he did not get it from a source he deemed reliable[3], but when the first issue of "Scientific American" arrived at his office that carried the news, the RPTV instantly reacted and asked Bell to send a pair of telephones. By chance, a pair of telephones did already arrive from an English acquaintance of Stephan on October 24, 1877. After five days of trials to determine the maximum range of transmission, Stephan decided to use the new device as an extension of the existing telegraph network for areas with small telegram income. The arguments in favor of the telephone were convincing: The price of a telephone was 1/80th that of a telegraph set, and its operators needed no lengthy and costly training. By a series of letters and of demonstrations to the Chancellor and the Emperor, the Post Office then initiated a process of consensus building. On November 28, 1877, the telephone was officially adopted as a further type of telegraph apparatus to expand the telegraph system into suburban and rural areas[4].

In Bavaria and Württemberg, the same experiments were run, with entirely different evaluations. In both states, the adoption was retarded for several years. In Bavaria, the telephone was tried as a replacement of the railway service telegraph and as a replacement of the inter-urban telegraph lines. For both purposes, the transmission range was too small. The telephone in its present state was then claimed not to be useful[5].

Besides the noted objective influences, the power of influential personalities in a time of small political elites and of political parties which were only in the process of institutionalization is obviously important. The swift introduction of the telephone into the German state telegraph system can also be attributed to the energetic and prudent personality of Heinrich Stephan, the head of the Post Office.

The second stage, the introduction as a subscriber network, started two years later. Already in October, 1877, after one week of experiments only, Stephan had the idea to make the telephone available to every household or business, an idea he repeated again in early 1878[6]. The idea was not implemented as Stephan found most of the prospective customers not to be interested at all. At the same time, a certain number of private point-to-point lines were built for private in-house conversations and for the internal communication within enterprises, public administrations, etc. even before the inauguration of the first state subscriber networks.

Compared to the introduction of the telephone into the telegraph system, where the telephone satisfied an existing need of a single customer who was also the operator of the system, the situation in

1880 was different. The operator and the customer were no longer the same entities, so that the operating agency had to consider the interests of its customers. For a business customer the existing services such as the mail service and the telegraph were operated at a reasonable speed and at moderate prices. Thus a contemporary would not perceive any "need" that had to be satisfied by a new technology. To make people use the innovation, a need had first to be created. In fact, this need was created by the RPTV, not by some purposeful action, but as an unintended effect of the performance of the telephone: The time-savings for those businesses that were already connected to the system forced commercial non-users to subscribe too in order to neutralize the communications advantage of their competitors.

The situation changed entirely when in 1880 the International Bell Telephone Company entered the scene. Because Bell's invention was not protected by a German patent, in Germany International Bell did not try to become a producer of telephone sets but instead intended to make its profits as a network operator. Thus the strong impetus to act came from outside the German communications administration. The application for private concessions first by Emil Rathenau, then by International Bell - the latter was even supported by the influential private banker of Chancellor Bismarck, Gerson von Bleichröder - and finally by several other private entrepreneurs, forced the Post Office to change its passive mood[7].

After a discussion within the Post Office between supporters of a policy of state concessions, who declared the telephone to be technically immature and therefore incompatible with the technically more sophisticated system of the telegraph, and backers of state intervention, who stressed the threat to the Reich finance and the danger of a loss of political and economic control to a foreign company[8], Stephan decided to interpret the legal situation of the telephone as being part of the existing state monopoly on telegraphy that was fixed by the Constitution. As a consequence of this decision, the RPTV was obliged to run telephone networks itself. After a lengthy search for subscribers, the first trial installation of a RPTV telephone exchange was opened in the capital, Berlin, on January 12, 1881, with eight subscribers only. On January 24, 1881, the first state telephone network was officially opened in Mulhouse (Alsace)[9].

In the meantime, Stephan asked Reich Chancellor von Bismarck to publicly support him. His intention was to produce a *fait accompli* in terms of legal decisions and technical facts because the legal interpretation of the Constitution by the Post Office was doubted. After Stephan made the Reich Treasurer support his demand, on February 12, 1881, the Chancellor publicly declared the telephone to fall under the provision of the Constitution for the state monopoly of telegraphy[10]. The decision can be interpreted as an example of the growing mood for state intervention in economic affairs among German politicians that started at the end of the 1870s.

The way the telephone was introduced in Germany heavily influenced its further development. The integration into the existing communication systems, the weak legal base of the integration of the telephone into the state telegraph, the form of tariff regulation and its impact on early spatial growth can be deduced from the starting conditions.

The slow growth of the telephone system in its starting phase can partly be explained by its integration into the state communications system. The financial effects of the advance of the telephone and of the relative retreat of the telegraph had to be adjusted within the same budget. Thus, new local networks were only opened if a threshold level of usually 50 subscribers guaranteed a minimum revenue to the Reich. The same policy of stressing financial safety was applied by the Post Office when interurban lines were constructed. In this case, the new subscribers or the municipal councils of the cities connected not only had to guarantee a minimum income to pay for the running costs but they even had to neutralize the estimated losses in the telegraph service[11]. Smaller communities had difficulties raising the money. The delayed expansion in rural regions was partly a result of this deliberate spatial extension of the actor network (as far as funding was concerned). In a situation which was characterized by uncertainty, the Post Office decided to steer a course of safety at the expense of system growth.

A new division of functions emerged. The telephone replaced the telegraph because it was cheaper and more rapid. The latter was used only if the telephone could not be used because of the initially limited transmission range of the telephone or if a legal document had to be produced[12]. Nevertheless, the telegraph did not "die". "Dying" of a technology because of a low intensity of use is only possible if two conditions are met: Firstly, considerations of profitability must be directly tied to the decision about the survival of the technology. Secondly, the new and the old system have to perform identical functions[13]. Both conditions were not met in the case of the German state telegraph system. The reason for keeping the telegraph system was not an economic one but purely a political one: The German PTT wanted to give every citizen access to a fast communication system even if he or she had no telephone; that the functions were different has already been mentioned.

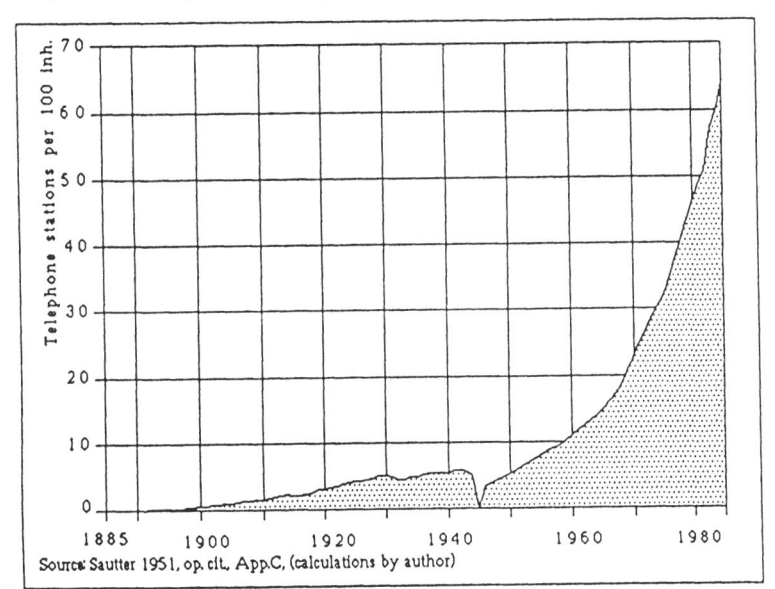

Figure 1: Development of Telephone Density, 1885 to 1985

Source: Sautter 1951, op. cit., App.C, (calculations by author)

The official definition of the telephone as a part of the existing state monopoly of telegraphs allowed the government to regulate the system by administrative decrees as it had done with the telegraph, i.e. without the consent of the Reichstag. This autonomous way of regulating extended the opportunity structure of the organization, though at the same time it became the cause of much trouble with political and economic actors who were left without influence. One of the consequences of this decision was the official definition of the function of the telephone in relation to the telegraph. The telegraph was perceived by the Post Office as the basic system for long-distance communication whereas the telephone was seen as an urban communications means, supplementing but not replacing the telegraph[14]. Given that strategy, in the first decades the Post Office tried to make the telephone grow with the help of its own revenue only. Only when it became apparent that interurban telephone calls in fact replaced telegrams and that the telephone system became profitable did the Reich Treasury allow the use of public loans for further growth.

The tariff that fitted into the early strategy was the flat tariff. It was the only form of tariff in which the revenue could safely be calculated in advance, and it needed no call-counting equipment. As an unintended consequence, the averaging of the subscriber rate that necessarily goes along with a flat tariff disadvantaged users in smaller networks that were not able to profit by the unlimited usability of the telephone that is part of a flat tariff.

The high flat rates (200 marks annually in 1881, 150 marks from 1884 to 1899) that resulted from that policy of self-financing severely limited the access to the system. Thus the first telephone subscribers were among those that relied on swift communications with only minor regard to costs: the information businesses, public and private administrations, the professions, and the well-to-do[15].

In line with the location and the geographical communication ranges of these first users, suburban, neighboring city and regional networks were established by the RPTV. The first interregional lines were built not as a network but as single lines, on a city-to-city basis, still in the style of the telegraph system. After 1887, with the rapid replacing of the telegraph by the telephone, the larger of the solitary local networks and their point-to-point intercity lines were step by step integrated into a nationwide network. A central reason for the pattern of spatial integration was the already mentioned lack of capital that favored a demand-oriented spatial growth. Therefore the geographical growth was not a development from chaos to structure[16] but a substitution process patterned by the spatial order of the pretelephonic communications space[17]. A second effect of the scarcity of funds was that it slowed down the spatial spreading of new technologies. Here, as well as during the construction of the long-distance cable network after 1921 or during the automation of long-distance switching after the Second World War, these communications technologies that all had the function to overcome distance in fact intensified spatial inequalities.

Due to the long initial phase of the process of spreading, the cities and urban agglomerations at the top of the central-place hierarchy increased their advantageous position.

The motive force of this first wave of spatial integration was an interaction between actors at three distinct levels. At the central state level it was the Post Office. There were regional actors, such as the chambers of commerce. They often cooperated with their political allies at the local level, in the city halls.

3 The legal stabilization of the telephone system

After a decade of unimpeded growth, the telephone system became so large that the casuistic solution of political problems characteristic of the initial phase had to be replaced by a more generalized way of solving conflicts with the political environment.

Up until the 1890s, the German telephone system consisted of scattered local and a few regional networks linked by a certain number of overhead wires. So the system essentially was an urban one. The first real obstacle to growth emerged with the growth of electrical utilities in urban areas. Both infrastructure systems used the ground beneath the city streets instead of building a special return line to save the costs for the return wire. Therefore the high-voltage lines were able to induce currents in the telephone lines that diminished the audibility of the conversations ("noise"). The telegraph administration was therefore interested in making the owners of high-voltage lines pay for the costly so-called "self-protection" of their facilities.

For the first time, the RPTV met with organized resistance to its plans. The municipalities supported the standpoint of the electrical utilities to which many of them were closely linked. Both actor groups were not at all interested in higher construction costs for power lines which would have been the result of the RPTV's policy. The fledgling electrical industry as a supplier of high-voltage equipment became another powerful ally of the utilities in the emerging fight. All of the opponents of the RPTV had strong supporters in the Reichstag, in the press, and in numerous "Electrical Associations" in the country[18].

The two primary intentions of the Post Office were to get a safe investment climate for the expensive cabling of the growing urban overhead wires and to prevent any competition for its subscriber networks to arise. The legal ground for the Post Office's intentions were not very strong. The municipalities to which the street ground belonged were self-governed (at least in Prussia) so that the RPTV was not able to enforce its view. In this situation the 1881 official definition of telephony as a part of the state monopoly of telegraphy proved to be a disadvantage. Even the highest German court had decided that a legal basis was missing to make the utilities and the municipal administrations comply with the wishes of the RPTV. Against its intentions, the telegraph administration was forced to give up the large freedom of action that until now the management by administrative decrees had given it. It had to accept that the regality of the telegraph and the telephone were to be fixed by a special law. This law could only restrict its range of action[19].

In general, the Telegraph Act of 1892 legally confirmed the status quo. The law legalized the state monopoly of the telephone, and it excluded all types of competition from networks with access for public users. Non-state (i.e. municipal or private) networks were confined to regions that the RPTV thought to be unimportant. In the case of the existing telephone networks of railways and of large enterprises, these private networks continued to be restricted to internal communication so that they could not compete with the state network. The conflict about the costs for the protection of the telephone wires against electrical induction was regulated by a compromise. Before, interactions between the RPTV and its subscribers were more or less regulated by private law. From then on and until today, they are regulated by public law which means a more institutionalized way of behavior on both sides. The Telegraph Act also extended the actor network because from then on, any raising of tariffs had to be passed by the Reichstag. With the entry of the Reichstag into the set of relevant corporate actors, pressure groups that had representatives in the Reichstag were now able to voice their approval or dissent, which they frequently did during the annual readings of the budget[20]. The political environment of the RPTV became more influential with the help of this law.

A comparison with the Prussian Law on Secondary Railways (Nebenbahngesetz) that was voted in the same year shows that the strategies of network control differed greatly between the telephone and the railways. Private secondary railway lines were allowed in areas situated between the great trunk lines of the Prussian State Railways. Their tracks were allowed to be connected to the state network[21]. The effect was a system of mutually supplementing networks. One reason for this way of regulating was that in Prussia private railway companies had a powerful tradition. Furthermore, investment costs for railway lines were far higher than those for telephone lines so that the Prussian government was not interested in extending its railway network into areas where it expected small returns only. The railway law increased the attractiveness of the state network and, at the same time, retained the state monopoly for the important part of the railway network. In contrast, network expansion in the telecommunications area was retarded until the RPTV itself had the financial power to serve the remote (and financially most unrewarding) rural areas.

A most important effect of the Telegraph Act for the expansion of the telephone system was that it increased the already existing incentive for the RPTV to continue extending its network into previously unserviced (i.e. rural) areas. From then on, its growth policy was split up between a demand-oriented variant as far as the establishment of new local networks was concerned and a supply-oriented strategy in the case of linking small communities to the long-distance network by public telephone stations. The reason for this differentiation of policy was that the RPTV had to operate a local network itself in order to prohibit the establishment of a competing municipal network and that a public telephone station was interpreted as a "network". Large numbers of rural communities had been connected to the telegraph system by the use of simple telephone lines between an auxiliary telegraph station and a full-service telegraph station at the nearby market center as a consequence of the decision of 1877. These internal telephone lines were then opened to the public. The RPTV thus averted criticism by stressing its performance in terms of geographical accessibility of the network.

The differentiated expansion policy was also an effect of the slow speed of expansion in combination with the size the system had then reached. In the initial phase of its geographical spreading the demand-oriented, hierarchical spatial growth of the interurban wire network connected only the larger among the German cities. Businesses in these cities that were enabled to use the telephone achieved a time-advantage in commercial communication over competitors in places that had no access to the interurban telephone lines and were restricted to the use of the slower telegraph. The representatives of these places

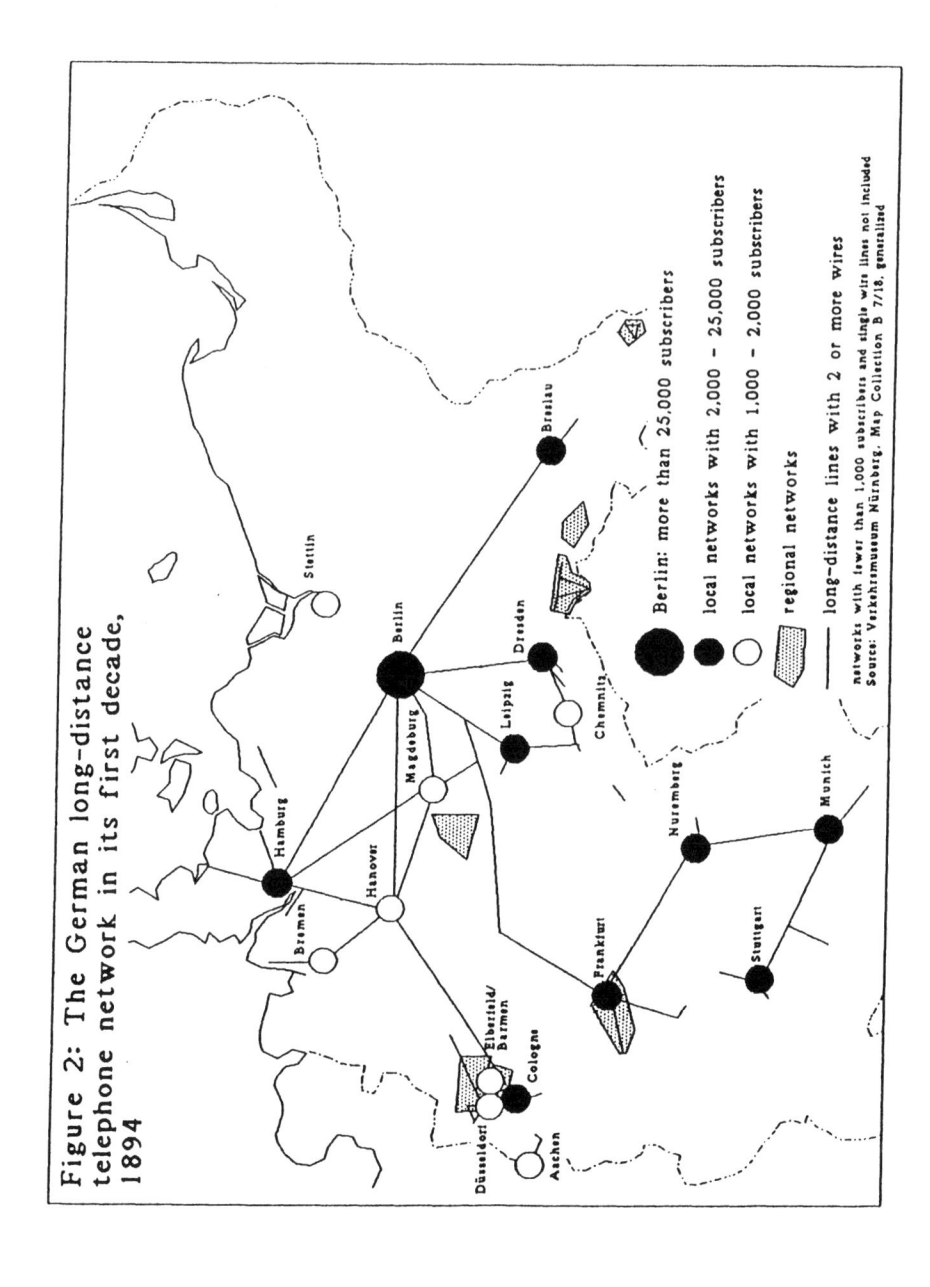

Figure 2: The German long-distance telephone network in its first decade, 1894

began to realize that being connected to the telegraph system only might become a problem and voiced their disapproval of the expansion policy of the RPTV[22]. Public telephone stations were a means to counter discontent; another, and more effective one, was the reduction of tariffs for small networks after 1900.

Within a few years, the solution to the growth problems in urban areas by the Telegraph Act produced new problems. Overhead wires in the cities were doubled to provide for induction-safe loop circuits. This, together with the continuous increase in subscriber stations, made the RPTV replace the overhead wires by expensive underground cables. At the same time, after 1887, the first generation of long-distance lines began to penetrate the countryside. The location of the major conflict shifted from urban to rural regions. The network no longer exclusively linked the large cities but the RPTV began to connect the rural market towns and county seats as well. This policy was impeded by smaller communities that had no interest in accepting the telephone lines along the roads that they had to maintain as long as they were bypassed by the network. In these situations, the RPTV thought it needed an all-German, unitary right-of-way (that did not exist) to assure the security of its investments. Therefore the intention of the RPTV was to harmonize its national action space with a corresponding "legal space".

The Telegraph Lines Act of 1899 extended the opportunity structure of the Post Office. The property rights of the owners of roads and of real estate had to be taken into account, but the institutionalized way in which these interests were now introduced into the planning process channelled their influence. The construction of future long-distance lines could no longer be severely impeded by individual or local interests. Again the public telecommunications system was made more powerful and more rigid.

Compared with the French development, the RPTV speeded up the spatial expansion of the network by centralizing the decision about the establishment of new local networks at the central state level by reducing the influence of local or regional actors on matters of right-of-way. In France, at the same time, the incorporation of local assemblies into the actor structure - as far as funding was concerned - retarded the spatial expansion[23].

The two laws of 1892 and 1899 were fundamental for the technical configuration of the telephone system. Their main provisions regulate the German telephone system still today. In stabilizing the interaction between RPTV on one side and its political environment and the subscribers on the other, the laws supported a stable growth rate until the First World War. With growing size, and thus, growing usefulness to the user, the system then became profitable[24].

The First World War had two contradictory effects on the development of the telephone system. It retarded the growth of the system, whereas the technological development was spurred. The war demonstrated the importance of a reliable long-distance network for the survival of the political system, and it changed the actor structure again.

The German war preparations foresaw a short war only. Therefore, the German telecommunications administrations went technically unprepared into the long war that finally emerged. In the summer of 1914, the second generation of long-distance equipment consisting of Pupin coils and amplifying valves just started to be put into practice. With the start of the war, a limited number of lines were handed over to the military for their exclusive use. For the first time in the history of the German telephone system, regular mass telephone traffic over distances of up to several thousand kilometers had to be realized. In theory, the problems could be solved, but in practice a war-induced scarcity of maintenance personnel and of imported raw materials decreased long-distance transmission quality. This led to a reappraisal of the more reliable telegraph for long-distance communications. The organizational side of the mobilization of the telecommunications system was the source of an enduring conflict. Because of the organizational autonomy of the RPTV on German territory, the military was not able to control the telephone system as a whole. So it tried to incorporate more and more long-distance lines into its separate network. A hidden struggle over the control of the network emerged. The beginning of the total war in 1916 meant a thorough reorganization of the whole military and civil telecommunications systems. The civil telephone network, although still run by postal personnel, was more or less exclusively used by military and war industry bureaucracies[25].

Another major effect of the war was the emergence of the trade unions as a new, important actor. Postal trade unions had already been founded at the turn of the century, but only when German labor was thoroughly reorganized for warfare, beginning in 1916, the unions got an accepted voice. For a short time, the telephone personnel exerted considerable influence through political strikes during the riots of

1919. The strike of leftists among the exchange personnel that cut the central state authorities in Berlin off from the entire nation could only be bypassed by changing from telephonic to telegraphic message transfer with the help of a loyal military telegraph unit[26].

4 Financial autonomy and modernization

The German defeat in the First World War changed the political environment of the RPTV. A new parliamentarian constitution made the Reichstag a core political actor. To compensate the loss of political power in foreign affairs, the Constitution favored further political centralization. In this connection also, the two southern German communications administrations were merged with the RPTV.

Civil war, demobilization, and inflation laid heavy burdens on the communications system. The whole interlocking communications system nearly stalled when the railway mail system broke down in 1919/1920 while the necessities to communicate quickly were higher than ever before (cf. Figure 5). Attempts were made to solve the problem of overload by expanding the supply of lines through a new underground long-distance cable network, through the temporary use of carrier frequencies on overhead wires and by rearranging the transmission capacity by new queuing rules. Replacing the unreliable overhead wire long-distance network by a cable network was a technical step towards national integration, paralleled in the transport sector by the merger of *Länder* railways into a nationalized Deutsche Reichsbahn and the planning towards a nationalized high-voltage power network spanning the whole of Germany. These movements, however, were not unique to Germany[27].

As a result of the new Constitution, every important change in the statutory structure of the telecommunications system now had to be regulated by law. Through the involvement of the Reichstag and the Reichsrat and of a special Traffic Advisory Council, the amendment of ordinances that now had to be passed by Parliament became too slow to cope with the quickly changing situation, especially in tariff matters[28]. The lack of maintenance during the war, overstaffing, and the large sums of money needed for the modernization of the system made the situation even worse. For the first time the RPTV needed to be subsidized. To support the German currency reform and to relieve the Reich budget of the burdens, the Reichspost Budget Law (Reichspostfinanzgesetz) was passed in 1924. This law changed the statutory structure of the German PTT in a fundamental way. It separated the property of the former RPTV from the Reich property while preserving the legal status of the Deutsche Reichspost (DRP), as it was now called, as part of the national administration. For the first time, DRP expenditures had to be balanced by revenues. To achieve this, the office heads were allowed to act in a businesslike manner, although still within the framework of the Reich Budget Law and the Civil Service Law. The Post Minister was charged both with the political control and the operational management of the system. A move of the Reichsrat (Chamber of *Länder*) to separate management and political control in the same way as it had been done in the Law on the Deutsche Reichsbahn one month before, failed.

Next to the Constitution, the Reichspost Budget Law is the basic *statutory* law of the whole postal and telecommunications history in Germany. It added economic interest organizations and the Reich Finance Minister to the relevant actor network by instituting a special Administrative Council. The power of the Council was centered on the control of the financial behavior of the DRP; it consisted of representatives of the Reichstag, of the Reichsrat, of the Reich Finance Minister, of the postal personnel, and of the organized business community. Regality matters continued to be regulated by the legislative.

Successive attempts to change the law or to abolish it altogether failed. The SPD and the Hansa-Bund[29], a business organization, which tried to reintegrate the DRP fully into the body of state administration, were as unsuccessful as the German Association of Chambers of Commerce and Industry and a transport business organization which both tried to convert the DRP into a stockholding company[30]. A structural change was not possible because all of the actors mentioned that were interested in such a change were able to obstruct one another. For the same reason, the most important organizational change within the German telephone system that was ever attempted by an outside actor, a lease by I.T.T. in 1931, did not have the slightest chance[31].

The management of the DRP was least interested in any change. The law of 1924 gave it sufficient scope for action. Interference by economic actors could be blocked by stressing the public function of the DRP. If on the other hand pressure from political actors became too strong, the DRP called attention to the necessity to manage its affairs in a businesslike manner[32].

In 1927, the Telegraph Law of 1892 was amended and given the new name of Telecommunications Installations Act (Fernmeldeanlagengesetz). Parliament further consolidated the state monopoly on telephony: it abolished the last legal basis for non-DRP networks *with public access* to be operated. For the first time, the military was explicitly named as an owner of monopoly rights. From then on, the German telephone system was split up into a network owned and run by the PTT with access to and for everyone, and a limited number of mutually unconnected smaller networks usually owned and run by large enterprises or public authorities for their own internal communications. The Telecommunications Installations Act is the basic law of German telecommunications in force until now.

A problematic effect of the Budget Law on the telephone system was that it sharply limited the financial maneuverability of the DRP at the very time when the telephone system needed large amounts of capital to expand (new telephone stations) and to modernize its technical facilities (automation of local exchanges, laying of the great underground long-distance cable network). The imposed financial constraints were subsequently eased by:

- cost accounting,
- lowering the running costs through rationalization,
- increasing both the accessibility and the usefulness of the system for the user,
- changing the structure of the telephone tariff and
- raising credits.

Cost accounting was very much needed because nobody really knew how profitable different ways of running the system were[33]. As there were no comparative data, thresholds of economic feasibility were fixed quite arbitrarily and tended to maximize economic safety. To rationalize, changes in the allocation of the personnel manning the exchanges according to test averages were introduced. Besides, personnel needs were reduced through the automation of local telephone exchanges. In rural areas this had a double effect: the number of operators was reduced and, at the same time, the service hours and thus accessibility for the user were extended. In fact, automating urban as well as rural exchanges had started before the war, but the speed of conversion was highest in the years before the world economic crisis of 1929[34]. The automation of long-distance switching was tested in that period in Bavaria for the same reason[35]. Replacing overhead wires by more reliable long-distance cables and enabling interregional and international calls by amplifying those calls increased the incentive to use those lines which were most profitable to the Reichspost. The unusually large number of five tariff reforms between 1923 and 1933 reflected the amount of outside pressure within the Administrative Council. Thus the tariff structure oscillated between a structure favoring large users and one encouraging households and other small users to subscribe. The Post Ministry closely cooperated with the representatives of trade and commerce to orient the telephone tariffs according to prime costs, i.e. it discriminated against the small user. The adversaries of this policy were those among the political parties that supported rural, private or small business users[36]. Load considerations became effective with the introduction of nighttime tariff reductions. Finally, the use of public loans was made easier, so that investments like the automation of exchanges or the construction of the cable network could now be financed by credits[38].

A contradictory effect of the new financial autonomy of the DRP was that it helped to modernize the existing system, but at the same time it impeded the growth of the number of private lines. The Reichspost was not interested in an unlimited growth of the system: Most of the newly connected subscribers in private households or small businesses did not use the telephone enough to make the extension profitable for the operator. In 1931, during the world economic crisis, the Ministry tried to remedy the situation by deliberately prohibiting its regional offices to advertise for new subscribers[39]. Another important and negative effect of the autonomy was that it induced the Reichspost to concentrate its efforts on the long-distance service in regions where return on investment was highest. During that period, long-distance cables connected only the larger local networks, whereas smaller networks were still linked by the less reliable overhead wires. Thus, the advantage of location which the cities of highest centrality already enjoyed was reinforced.

The Reichspost Budget Law stabilized the public communications system. It established a certain degree of financial and managerial autonomy. Because of the lack of competition the DRP was able to concentrate its efforts on the technical improvement of the existing system and neglect the existing demand. The idea that the usefulness of the system was increased if everyone had access to it and that usefulness to the user meant more revenue to its owner was never embraced by the DRP in that period.

5 The telephone system under military control

The period of Nazi reign between 1933 and 1945 shows the most dramatic change in the function and the actor configuration of the German telecommunications system before its expansion after the Second World War. More than ever before, the function of the system became a political one. The NSDAP, following its strategy to establish "the unity of party and state", used the telecommunications system as a means for the control of the German population and, after a period of preparation, for the war that was intended to create a Greater Germany. The German military, in implementing this policy, expanded the long-distance telephone and telex system at a speed that had never before been attained. It is this aspect of the political function of the telephone system which this section aims to elaborate.

The period of Nazi control can be divided into three consecutive stages: In the beginning, the telephone system was brought under the exclusive control of the political system. After 1935, it was prepared for war. Finally, it was used for that purpose.

The NSDAP seized power within the DRP early in 1933. First the function of the communications system was changed. Before this time, the DRP had defined itself as "a servant to communications", i.e. as an infrastructure system for both the economy and the government. Now the major function of the communications system was shifted to make it into a means for political power, to serve as a command-and-control system. For the growth of the system it meant that cost considerations were replaced by an infrastructure approach.

In 1934, a move to abolish entirely the organizational autonomy of the DRP failed[40]. The organizational "momentum" of the Reichspost had become too large to be dissolved without resistance. On the contrary, what remained of this initiative meant a centralization of decision-making by excluding pressure coming both from the environment (economic and regional political actors) and from within the organization (personnel). The Administrative Council was abolished as well as the last remnants of separate regional organizations in Bavaria and in Württemberg[40].

A staunch follower of Hitler, Wilhelm Ohnesorge, was made Secretary of State of the Postal Ministry[41]. The personnel was purged at once; 10% of all senior officials had to leave the Reichspost. In the lower ranks nearly three thousand new employees entered the service in the first year, most of them formerly unemployed SA- and SS-men. The party influence was increased by selectively promoting party members[42] and, at least after 1937, by compulsory NSDAP membership of newly appointed senior officials[43]. Since 1933, every chief official of each of the regional administrations was controlled by a special official who directly reported to the Minister[44]. A Post Militia (Postschutz) was established to protect telecommunications and radio facilities in times of political unrest or war. Its peacetime function was to intimidate the non-party members among the DRP personnel.

Very soon, the Gestapo widely controlled telephone conversations, with the necessary technical help of DRP experts[45]. The policy of the NSDAP thus openly showed the *double function of support and control* that the state monopoly of telecommunications had: to enable people to exchange their ideas and, at the same time, to control that exchange if it becomes a threat to the political system. So the same Janus-faced function the mail system always had was ascribed to the telecommunications system.

The legal and organizational preparations for war began in 1935. In that year, an amendment of the Telegraph Lines Act helped to keep secret the extension of the cable network that the Wehrmacht was to finance. The law strengthened the position of the DRP in matters of rights-of-way so much that today, when the reason for its existence has since long disappeared, the law is still in use[46].

In this period, the Nazi government used a double strategy. The DRP was led by senior officials who were party members, if not convinced NS-followers, but who were also DRP-men. So the interests of both sides, the build-up of military strength and the traditional interests of the DRP to run a nationwide system that is technically up-to-date and performs well, were pursued at the same time.

In 1937, a special Communications Department within the Supreme Military Command was established, with branch offices at every regional Reichspost administration, to control the functional integration of the civil telecommunications organization into the military in the event of war. In 1938, the (unpublished) Reich Defense Act placed the Reichspost in time of war under the direct command of the Supreme Military Command as it did the two state traffic organizations of the Deutsche Reichsbahn and the Autobahn[48]. The military did everything to avoid repeating the bad experience of the First World War. Its behavior shows the pattern of indirect rule exercised during the following years. The core interest of the DRP, its survival as an organization of its own, was safeguarded, but political indoctrination by the Nazi party and military control of the access to the scarcest of the resources, expert manpower and raw materials, enabled the military to use the DRP to the fullest.

The technical preparations for war began in 1936. If the Wehrmacht wanted its control-and-command communications to survive a future air war and still fight a mobile war, a communications network with a high amount of redundancy and invulnerability against aerial attacks was needed. The already existing public telephone and telegraph cable network was one of the most extended ones in Europe. Therefore the most economical way to improve military strength in the communications sector was to divert its use in time of war to the military and not to build a separate network. The first Four-Year-Plan issued in 1936 gave the Reichspost the necessary financial resources, so it did not find it too difficult to comply with this scheme[49].

The Reichspost was made to expand and to modernize its network. The existing star-shaped underground cable network had to be changed into mesh form and to be extended into rural areas in western and southern Germany. These areas had been bypassed by the first cable network because of the demand-oriented and cost-sensitive expansion policy of the DRP at that time but after 1935, army garrisons or air stations were located in these regions, especially in western Germany (Westwall). Long-distance cables had to be laid along secondary roads to reduce the chance of being hit. The most important network nodes Berlin, Frankfurt, Munich and the Ruhr got bypass cables. Repeater facilities were moved to sites outside the city centers, mobile repeater units constructed and emergency long-distance exchanges were built under the cover of air-raid shelters[50]. After the war, sheltered exchanges often were the only ones that remained intact to reestablish telecommunication links[51].

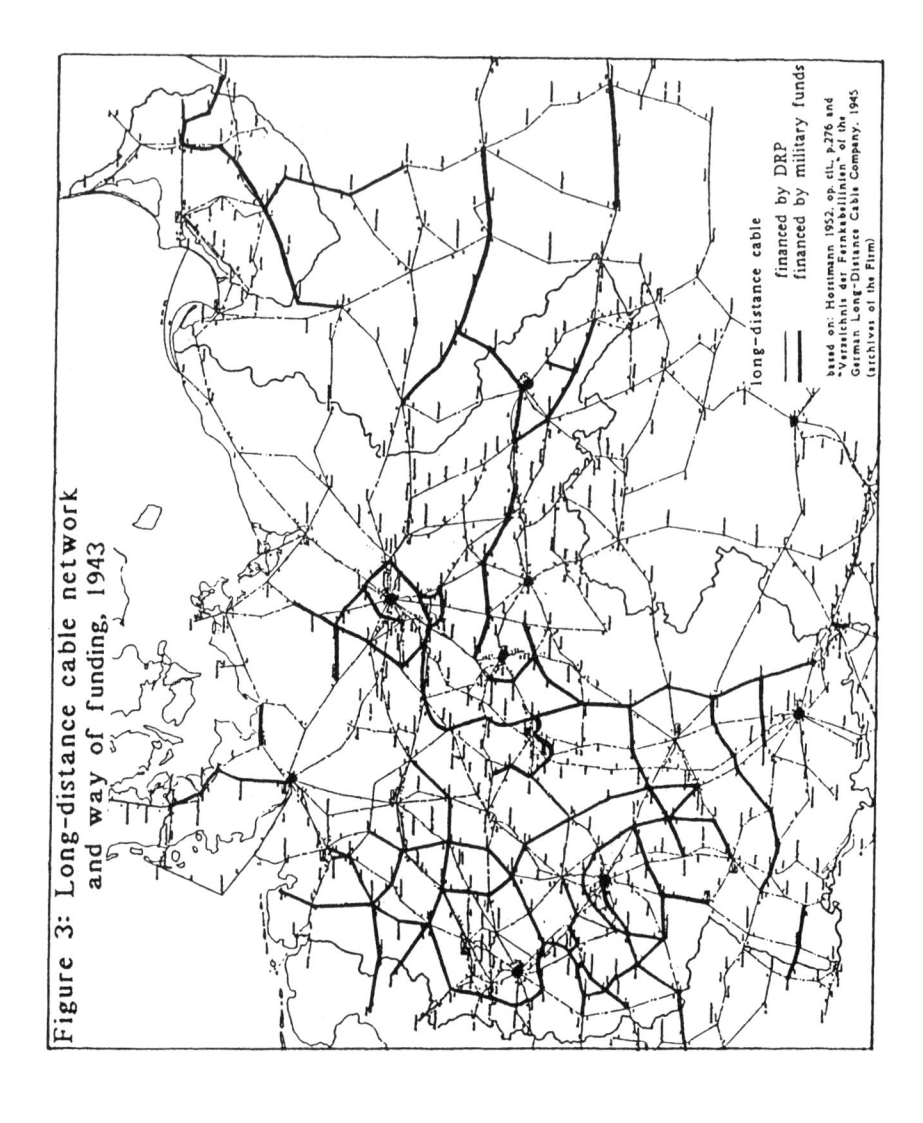

Figure 3: Long-distance cable network and way of funding, 1943

Modernization was a second feature of strengthening the telephone system. Coaxial cables connected the centers of political and military power (Berlin, Nuremberg, Munich, Hamburg, Frankfurt, Vienna, Hanover). In peacetime, the coaxial lines permitted economies of scope superior to the old standard telephone and telegraph cables: The technical function of the lines was to carry the centrally produced TV pictures to the first TV transmitters, to enlarge the number of circuits in telephone and telex transmission by enabling the use of carrier frequencies on cables and to transmit videophone calls. The star-shaped pattern resulting from the location of the TV production center in Berlin reinforced the locational advantage of the capital. In local networks, the new service of high-frequency wire broadcasting on telephone lines enabled a higher quality of broadcasting than on radio waves and, after 1939, to curb listening to enemy radio stations and to transmit detailed air-raid warnings without being monitored by Allied forces[51].

After the start of the Second World War, the military exploitation of the telephone system proved to be well planned. The Supreme Military Command became the core actor and controlled the telecommunications system with the help of the DRP experts. The military took full advantage of the system: On some cables more than 50% of the circuits were handed over to the military[52].

As the war went on and Hitler subjugated the majority of European countries, a genuine telecommunications *Geopolitik* emerged step by step. All of the networks in German-occupied countries were incorporated into the German one to serve as a unified Wehrmacht command-and-control system[53]. In October, 1942, the convention of the European Postal and Telecommunications Union was signed in Vienna, integrating the respective administrations of the German-dominated countries[54]. Special telecommunications attachés at the German embassies were to reinforce organizational links.

After the start of the Allied combined bomber offensive in mid-1943, the telecommunications system gradually broke down. The action pattern of the DRP changed from planned behavior to mere improvisation as the accumulating amount of destruction became overwhelming. But the way in which the DRP coped with the losses which the Allied attacks inflicted on its system highlights once more the political function of the telephone system. To counter the effects of Allied bombings, a choice of alternatives existed for the DRP. These alternatives were related to:

- the technology of transmission (radio or wire),
- the transmission range (local or long-distance),
- the organization of maintenance (centralized or decentralized),
- the time sequence of usage (private, business, official or military).

As the primary function of the telecommunications system was the survival of the political system, the repair work emphasized the maintenance of the long-distance cable network that was necessary as a means for political integration. A special maintenance organization for the areas hardest hit was set up in July, 1943. It was assisted by maintenance teams of the special mixed private-public enterprise that had laid the cable network and later even by military signals units. An official radio telecommunications overlay network was planned to increase redundancy that was lost due to the bombings[56]. Private and business users had to use the telegraph or the mail service. Overload was curbed by increasingly tight queuing rules for long-distance telephoning, by the possibility of shutting down entire regions and by disconnecting private users[57].

In this time of utmost pressure Wehrmacht and DRP cooperated closely. Both sides urgently needed more lines than those that were left in operation. Cooperation in technical maintenance was extended to cooperation in utilization: The military allowed official civilian users to use their lines in closely specified situations[58]. As a result of Allied bombing and German countermeasures, the spatial structure of the telephone network changed. After 1944, the physical existence of the technical infrastructure and its usability started increasingly to diverge. Nevertheless, up to the very end of combat, a telecommunications system at the strategic level, reduced as it was, continued to work[59].

6 Reconstruction and expansion of the system

When the Allied forces occupied all of Germany, all forms of telecommunication were forbidden by Allied Proclamation No. 76. By military decrees of local military governments and by the initiative of DRP personnel a slow reconstruction and a very limited service began in the summer of 1945. By this period, the integrating power of the DRP's organizational culture became clearly evident.

The war had destroyed neither the personnel nor the organization and the procedures. But the technical system had to be reconstructed, even if there were large parts that had survived intact. In the larger towns the technical infrastructure was most heavily destroyed so that the most modern equipment was most in need of repair. Also, the telecommunications industry was badly hit. Before the war, the largest telecommunications suppliers were all located in Berlin. Now, nearly all of them were bombed to the ground, and what remained was dismantled. Slowly, the enterprises began to reorganize in Western Germany. Though the material structures had mostly vanished, and initially German patents were no longer respected, the social structures of the firms, their relations with the telecommunications administration, and the personnel with its accumulated experience and expertise had not.

The Allied occupation did not alter the organizational structure of the telecommunications system too much, at least not in the long run. The local and regional level of the postal and telecommunications administration remained structurally unchanged, whereas the top ministerial level was reorganized according to the intentions of the different political powers. For a few years, there was a scarcity of experienced senior personnel because of the high numbers of former NSDAP party members among the experts that were temporarily ousted during denazification.

One of the earliest intentions of the Allied powers was to reconstruct the telephone network for their own needs. In all of the occupied zones, the reconstruction of technical and administrative structures of the telephone system was therefore among the first steps taken by the Allied postal and telecommunication officers. The very early establishment in September 1945 of an organization at the ministerial level for the British Zone, the Reichspost-Oberdirektorium, and of a consulting agency consisting of German officials for the whole of the French zone, can be traced back to the same purpose. At this time, the political control function of the telephone system remained active, but its economic function, including physical survival of the population, became more and more important[59].

Even before the currency reform of 1948 and the passing of a new constitution one year later, the combined postal and telecommunications administration for the British and U.S. occupied zones reestablished contact with the telecommunications industry. An Advisory Committee for Communications Technology was founded to coordinate the future construction of German telecommunications. The function of this committee was to develop technical norms. In the spring of 1948, it was decided to automate long-distance switching, to change the switching technology and, as a prerequisite, to lay a new carrier frequency long-distance cable network. The geographical structure of the network had to be adapted to the new political and economic geography as well[60].

One reason for making such far-reaching decisions was surely that the war had only interrupted a development that had already begun. Full-scale planning to adopt automated switching (although limited to regional districts) had already started during the 1930s[61], and in 1938, Siemens & Halske had built a forerunner of the rotary switching technology that was used after 1955. With the internal telephone system of the Deutsche Reichsbahn, a decade-long experience with a nationwide dialing system - if again of limited complexity - existed[62]. Another reason is probably the continuity among technical experts and senior officials. Dr. Steidle, for instance, who was responsible for the first large-scale experiments made in Bavaria in the 1920s, reinitiated planning measures already in 1946[63] and later headed the central research and development agency of the German PTT. Moreover, the heavy destruction of the telephone system was perceived by the industry as a chance to construct an advanced system. This accorded well with its traditional export strategy. The German telecommunications industry welcomed state commissions as an opportunity to demonstrate its regained technological modernity to potential buyers abroad[64]. In any case, the decision to modernize was a deliberate step towards an infrastructure approach in network policy: The telecommunications administration intended to take its part in the reconstruction of the devastated country.

The technical norms for the new network were set at the 1946 and 1949 conferences of the CCIF, the International Consultive Committee on Telephony, a suborganization of the International Telecommunications Organization, without German participation. The Committee agreed on a network of carrier frequency lines as the new backbone of European international communication[65]. If the German PTT wanted to exploit the geographic advantage of Germany's position in the center of Europe and to attract transit traffic again, its long-distance network had to meet the aforementioned international standards. The main lines for the new network in Germany were then laid between 1949 and 1956.

New cables were developed that were adapted to carrier frequency use, and the use of styroflex plastics instead of copper for cable production reduced capital needs for transmission lines. Automated switching - although capital-intensive - and the incorporation and extension of former military multiplex microwave lines further reduced running costs. As in France, international military funds helped to finance some of the major lines[66].

After the foundation of the Federal Republic, a Federal Ministry of Posts and Telecommunications was established and the organization changed its name into Deutsche Bundespost. In 1953, the Bundestag passed a new PTT Administrations Act that, in most of its fundamental provisions, resembles the Reichspost Budget Law of 1924. Financial autonomy was confirmed, though the links of the Bundespost to the national bureaucracy were strengthened.

The second major characteristic of the post-war period was the start of mass distribution. Already in 1951, the telephone density (telephone stations per 100 inhabitants) for West Germany surpassed the all-German level of 1938 (cf. Figure 1). The system changed its character from a business tool and a luxury for the few to a *mass system*. It is not this orientation of the German PTT in 1948 and 1949 that was surprising, but its timing in the face of the poor conditions of the economy. In fact, the Bundespost merely followed the infrastructure approach with respect to subscriber growth of its predecessor. Already in the 1930s, the DRP had induced a growth of telephone stations by tariff reduction and promoted the development of a low-cost dial party-line technology to meet the low revenue expected from future small users[67]. What remains astonishing is that after the war, in an economic situation worse than ever before, the technological standard of the telephone station technology was even higher. The close integration of the PTT into national administration that the Basic Law as well as the PTT Administrations Act confirmed might be one reason. As a consequence, the right of every applicant to a telephone with the same operational quality and an infrastructure approach were stressed. Aside from this, the reasons already pointed out in connection with long-distance automation may have played a role.

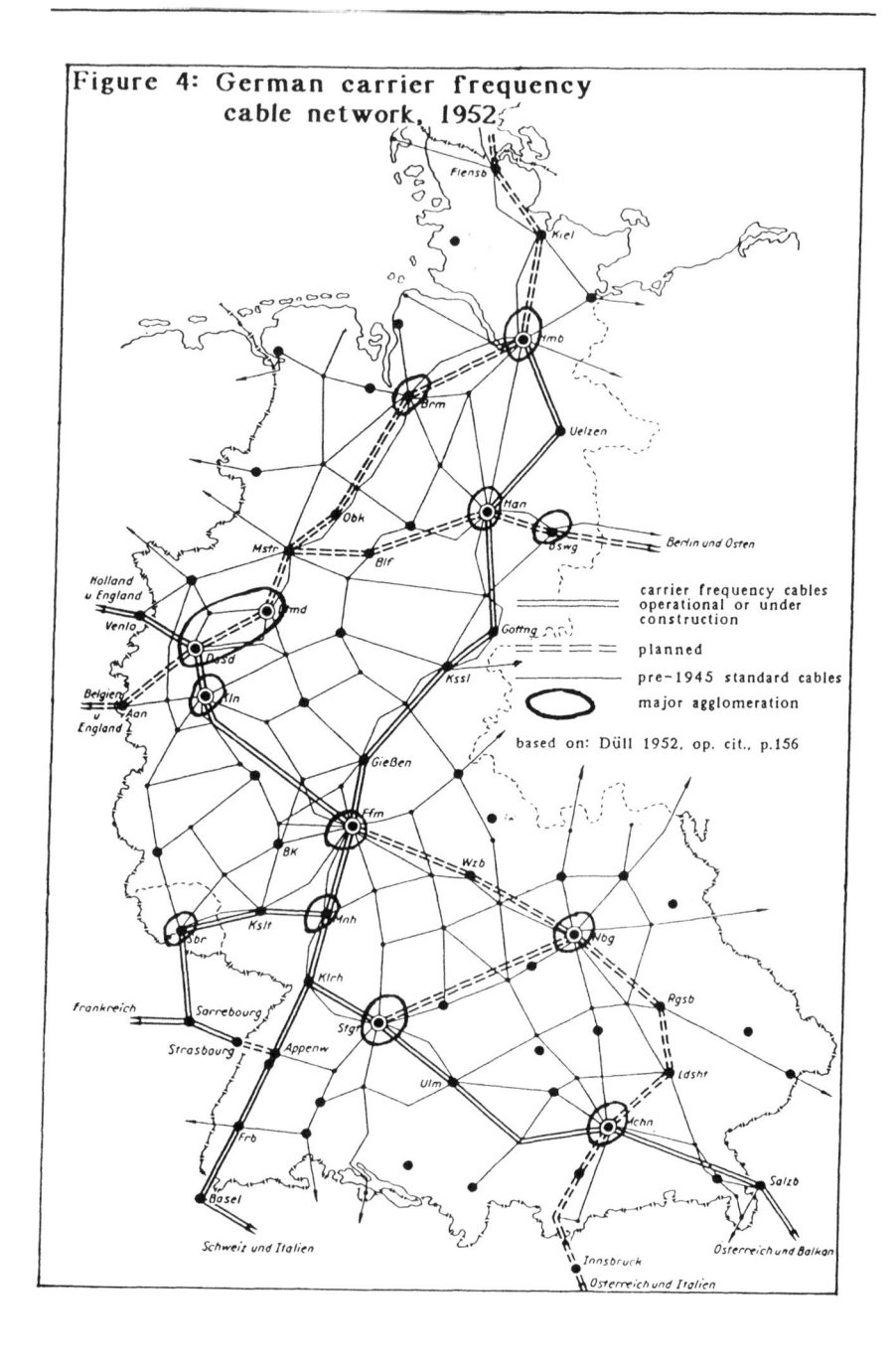

Figure 4: German carrier frequency cable network, 1952

Economically, the expansion of the telephone system was made possible by a mutual reinforcement of supply and demand. The supply of telephone services was increased at low prices. The charge for a local call was viewed by politicians as a so-called "political price": Any increase herein was thought of by the federal government as an indicator of what the public might understand as inflation[68]. Therefore the federal government let the tariffs stay at the same level from 1954 to 1964. The demand for new private lines was made possible by the continuous increase in the purchasing power of the average household. System growth accelerated after the mid-1960s when wages went up and the majority of households had satisfied their immediate needs of housing, clothing and better eating. Today only residual household groups do not have a telephone at home, but compared with countries of comparable GNP per capita that started earlier with their telephone mass distribution (such as Denmark or Sweden), the West German system in 1985 still lags approximately one decade behind[69].

The geographical structure of the new network reflects the new political and economic space structure as the old one before 1945 mirrored previous spatial hierarchies. The new political system lacks the strong central position Berlin once had. A new spatial dispersion of economic activities further favored the regional centers. The high capacity carrier frequency network combined a star-shaped network that was generated in the dispersed local networks with a ring network that channelled interregional traffic. At the same time, the network structure stresses the north-south axis of all traffic in Western Germany (cf. Figure 4). The prewar cable network and its former military extensions now served as regional feeder lines.

With the automation of long-distance switching, the use of the long-distance lines increased enormously. Three reasons can account for this. First, for the first time the long-distance network became really attractive to the user as it enabled virtually instant communication with distant partners. Secondly, the German "Wirtschaftswunder" propelled the economy into previously unknown heights. Lastly, by calculating the call charge on the basis of the time used instead of on a minimum time of three minutes as in the era of manual operation before, automated long-distance calls were cheaper than manually switched ones[70].

The heavy use of the long-distance network was not expected by the Bundespost. The spatial structure had been planned according to the use intensity and distribution of the operator-controlled era. After some time, the high use intensity made it necessary to link not only the highest levels, but even lower levels of the switching hierarchy with a mesh-form network.

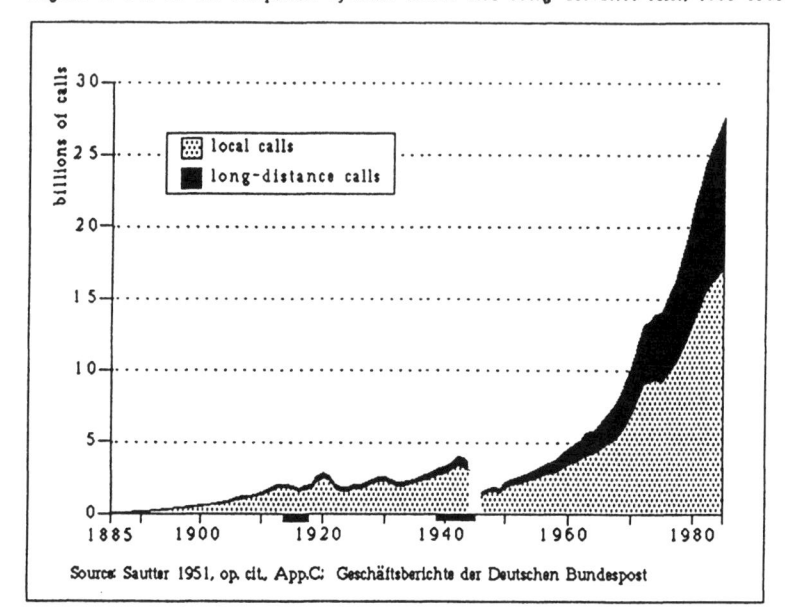

Figure 5: Use of the telephone system: Local and long-distance calls, 1885-1985

Source: Sautter 1951, op. cit. App.C; Geschäftsberichte der Deutschen Bundespost

7 Conclusion

A number of lessons concerning the development of large technical systems can be learned from the history of the German telephone network.

First and above all: The development of the system is driven by decisions of a limited number of actors. If a certain amount of momentum developed, this was not a result of forces inherent in an autonomous technology but of purposive action constrained by the sediment of previous decisions about technological alternatives.

Second: The decisions made at the start heavily influenced the course of development of the system in the long run.

Third: Because of the integration of the telephone into the state communication system, in Germany political actors had the say also on economic questions. Such integration often makes it difficult to distinguish between the telephone system and other state communication systems.

Fourth: Political, economic, social, technical and geographic aspects of system development became tightly interwoven.

Fifth: In the case of a large-scale communications system which aims at overcoming geographical distance it is worthwhile to stress the geographic differentiation of the actor system and of the geographic properties of the system itself.

8 Notes

BA   content of a file in the Bundesarchiv, Koblenz
RGBl.   Reichsgesetzblatt

1 Up to 75 km of operational range in the early years. Ernst Feyerabend, 50 Jahre Fernsprecher in Deutschland 1887-1927. ed. by Reichspostministerium, Berlin, 1927, p. 31.
2 Feyerabend 1927, op. cit., p. 27.
3 Paul Henseler, "Aus der Frühzeit des Telefons", Archiv für deutsche Postgeschichte, (1975)2, pp. 164-166.
4 Feyerabend 1927, op. cit., pp. 25-28.
5 Willy Feudel, "Zum 100jährigen Jubiläum des Telefons in Bayern", Archiv für Postgeschichte in Bayern (1983)1, pp. 3-4. At this time of research, no apparent reason can be given for the non-adoption in Württemberg.
6 Conrad Matschoß, Werner Siemens, Vol. 2. Berlin: Springer, 1916, p. 535.
7 Jürgen Schliewert, "100 Jahre Fernsprechen in Frankfurt am Main". Hessische Postgeschichte 26(1981), pp. 32-33; Heinz Wernick, Die Anfänge des Telefons. 100 Jahre Telefon im Wuppertal, ed. by Fernmeldeamt Wuppertal, Wuppertal 1982, p. 5.
8 According to the autobiography of one of the participants: Paul David Fischer, Erinnerungen aus meinem Leben. Berlin: Springer, 1916, pp. 230-231.

9 René Muller, "Mulhouse, premier réseau téléphonique d'Alsace ... et d'Allemagne (1881)". Cents ans de téléphone en Alsace = Diligence d'Alsace (1976)16, pp. 21-22.
10 Oskar Grosse, 40 Jahre Fernsprecher. Stephan - Siemens - Rathenau, Berlin: Springer, 1917, p. 31.
11 Billig, "Die Entwickelung des Fernsprechwesens im rheinisch-westfälischen Industriegebiete 1881 bis 1886". Archiv für Post und Telegraphie (1887)14, p. 429; (1887)15, p. 454.
12 After the break-through in the development of long-distance communication equipment in 1887, the share of the telegraph service on national, long-distance messages dropped by 50% within 13 years. Derived from data compiled by Karl Sautter, Geschichte der Deutschen Reichspost 1871 bis 1945. Geschichte der deutschen Post, Teil 3. Frankfurt/M: 1951, Appendix C.
13 Colin Cherry, World Communication: Threat or Promise. A Socio-Technical Approach. London, New York, Sydney, Toronto: Wiley-Interscience, 1978, p. 46.
14 Reply of the Post Office to several Chambers of Commerce that had complained about its prohibitive conditions on behalf of opening new inter-city lines. R. van der Borght, "Die Thätigkeit der deutschen Handelskammern in Bezug auf das Fernsprechwesen im Jahre 1889". Jahrbücher für Nationalökonomie und Statistik 56(1891), p. 423.
15 No overall historical sociology of telephone usage for Germany exists until now, though identical local examples are given by Elfriede Meurer, "100 Jahre Telefon in Köln 1881-1981", Mitteilungen der Gesellschaft für deutsche Postgeschichte Bezirksgruppe Köln, (1982)7, p. 34; and Heinrich Walters, "100 Jahre Fernsprechnetze in Minden und Münster", Postgeschichte in Westfalen (1987)4, pp. 88-89.
16 Heinze/Kill, this volume.
17 Borght 1891, op. cit., p. 422.
18 Author unknown, "Die Berathung des Entwurfs eines Gesetzes über das Telegraphenwesen des Deutschen Reichs im Reichstag". Archiv für Post und Telegraphie (1892)7, pp. 222, 232-233.
19 Hermann Wiltz, Das Gesetz über das Telegraphenwesen des Deutschen Reichs (Telegraphengesetz) vom 6. April 1892 nebst dem Gesetze, betr. Abänderung dieses Gesetzes (Telefunkennovelle) vom 7.März 1908, Frankfurt/M.: Wolstein & Teilhaber, 1908, p. 1.
20 As it was evident in the reports published year after year in the Archiv für Post und Telegraphie, a supplement of the post office's administrative bulletin.
21 Heinze/Kill, this volume.
22 Annual report of the Chamber of Commerce of Halle (Saxony) for 1889, cited by: Borght 1891, op. cit., p. 422.
23 Jacques Attali and Yves Stourdze, "The Birth of the Telephone System and Economic Crisis: The Slow Death of the Monologue in French Society" in Ithiel de Sola Pool (ed.), The Social Impact of the Telephone, Cambridge/Mass.: MIT Press, 1977, pp. 97-111, and Jean-Paul Martin, "Les premiers développements du téléphone en Lorraine (1885-1914)", Revue Géographique de l'Ést (1982)3-4, pp. 215-234.
24 Kurt Schubel, "Zur Geschichte der Finanzwirtschaft der Deutschen Reichspost- und Telegraphenverwaltung von 1871 bis 1918", Archiv für das Post- und Fernmeldewesen (1968)4, p. 422-423.
25 Grallert, "Das Telegraphen- und Fernsprechwesen vor, in und nach dem Kriege", Archiv für Post und Telegraphie (1921)1, pp. 13-14, 23-24.

26 Letter from the Postal Ministry to the Reich Chancellery re.: list of line break downs due to the Spartakist riots in March and April, 1919, dated May 28, 1919, BA R 43 I/1993 Bl. 9-15; and letter from the Office Workers Council of the Special Telegraph to the Under Secretary of State of the Reich Chancellery, dated January 3, 1921, BA R 43 I/2006, Bl. 39-42.
27 Cf. the volumes of "Europäischer Fernsprechdienst" that chronicle the growth of national long-distance cable networks in the 1920s all over the industrialized countries.
28 As a result of the galloping inflation, the telephone tariffs had to be altered once in 1919, in 1920, and in 1921 each, four times in 1922, 16 times in 1923, according to: Gottfried North, "Die Entwicklung der Fernsprechgebühren, in Deutschland", Archiv für deutsche Postgeschichte (1977)1, (Special issue: Hundert Jahre Fernsprecher in Deutschland), pp. 210-213.
29 Application of the Hansa Federation for Commerce, Trade, and Industry to the Reich government and the Reichstag re.: amendment of the Reichspost Budget Law, dated December 17, 1927, BA R 43 I/2006.
30 Post und Eisenbahn. Tatsachen und Gedanken zur Vereinheitlichung der deutschen Verkehrspolitik. Memorandum, Berlin: DIHT, 1932; Saliger, "Reichsverkehrsreform". Magazin der Wirtschaft 6(1930), pp. 1675-1676.
31 Correspondence between the representative of the Morgan Trust, Prince Löwenstein, the Reich Chancellery, the Reich Finance Ministry and the Postal Ministry, all dated between October 6, and November 9, 1931. BA R 43 I/1998 Bl. 361-374
32 Karl Sautter, "Organisationsfragen des deutschen Verkehrswesens", Archiv für Post und Telegraphie, (1932)6, pp. 145-149.
33 Schubel 1968, op. cit., p. 385.
34 Feyerabend 1927, op. cit., pp. 171-172.
35 Willy Feudel, "Die Entwicklung der Netzgruppentechnik in Bayern". Archiv für Postgeschichte in Bayern (1978)2, pp. 326-340.
36 Cf. minutes of the session of the Working Committee of the DRP Administrative Council on January 28 and 29 and February 2, 1927, BA R 48/207, pp. 2, 4.
37 Gebbe, "Aufbringung der Geldmittel für den Ausbau des deutschen Fernsprechnetzes". Europäischer Fernsprechdienst (1928)7, p. 60.
38 Ernst Feyerabend, Der Einfluß der Tarifpolitik auf die Entwicklung des Fernsprechwesens in Deutschland. Jahrbuch für Post und Telegraphie (1928/29) pp. 19-20,27; circular of Postal Ministry to all Reichspost regional administrations re.: advertisement in official telephone directories, dated May 15, 1930, BA R 48/312.
39 Letter from the Reich Ministry of the Interior to the Reich Chancellery dated February 15, 1934, BA R 43 II/1147, Bl. 13.
40 Law on the Simplification and Cost Reduction of the National Administration, issued February 27, 1934 (RGBl. I S.130).
41 He became Minister in 1937 and remained in office up to the very end of the war.
42 Minutes of the annual conference of DRP regional directors in the Postal Ministry at Berlin, on June 6, 1934, App. 2, BA R 48/238, pp. 3, 8-9.
43 Testimonial by Hans Schuberth, Director of Posts and Telecommunications for the American and British Zones of Occupation, dated October 20, 1947, printed in: Kurt Wiesemeyer, Das Personalwesen der Deutschen Post vom Zusammenbruch des Nationalsozialismus bis zu den Anfängen der Bundesrepublik, ed. by Bundesministerium für das Post- und Fernmeldewesen. Frankfurt, 1954, App. 1, pp. 156-160.

44 Appendix to the letter of the complaint of the retired DRP regional director Ringel to Hitler, dated December 8, 1937, BA R 43 II/1147 b Bl. 23, 36, 41.
45 Tapping devices were installed by the DRP, in some cases the DRP owned or even operated them, cf.: circular Reichsführer SS and Head of the German Police to all Gestapo-offices, dated August 28, 1937, BA R 58/242, Bl. 174.
46 Letter of the Reichspost Minister to the Head of the Reich Chancellery dated July 9, 1935, BA R 43 II/267.
47 Gerhard Meinck, "Der Reichsverteidigungsrat". Wehrwissenschaftliche Rundschau, 6(1956)8, pp. 420-422.
48 Hensger, Die Nutzung der Fernmeldenetze der Deutschen Reichspost durch die Wehrmacht im Zweiten Weltkrieg. Typescript, Feldafing, 1984, p. 4.
49 Walter Surén, "Erfahrungen von 1914 wurden im Zweiten Weltkrieg voll genutzt". Fernmelde-Impulse 6(1965)2, p. 55.
50 In the case of Frankfurt/M.: Schliewert 1981, op. cit., p. 61; Kiel: Willi Jensen, "100 Jahre Fernsprechen in Kiel. 1883/1983". Post- und Fernmeldegeschichte zwischen Nord- und Ostsee 23(1983)1, p. 334.
51 Gerhart Goebel, "Der Deutsche Rundfunk bis zum Inkrafttreten des Kopenhagener Wellenplans", Archiv für das Post- und Fernmeldewesen 2(1950)6, pp. 420, 438.
52 Circular of Reichs Post Minister to all supreme Reich administrations re.: to employment situation and work distribution in the DRP, dated February 19, 1944, BA R 2/21203, p. 2.
53 Hensger 1984, op. cit., pp. 6-9.
54 Werner Zschiesche, "Der Europäische Postkongreβ Wien 1942", Europäischer Fernsprechdienst (1942)61, pp. 92-93.
55 Letter of the Secretary of State of the Postal Ministry, Nagel, to Reich Minister Goebbels, Plenipotentiary for Total Warfare, dated January 3, 1945, BA R 43 II/638 Bl. 196r, 197.
56 Erwin Horstmann, 75 Jahre Fernsprecher in Deutschland 1877-1952, ed. by Bundesministerium für das Post- und Fernmeldewesen. [Frankfurt/M.] 1952, pp. 165-166.
57 Military regulation of the Head of the Communications Department within the Supreme Military Command re.: joint use of communication facilities of the Wehrmacht, dated June 17, 1944, BA R 43 II/481, Bl. 40-42.
58 Hensger 1984, op. cit., p. 9.
59 Ludwig Kämmerer, "Der Wiederaufbau der Post in der britischen Besatzungszone", Archiv für deutsche Postgeschichte (1978)2, pp. 10, 12; Wilhelm Ebenau, "Der Wiederaufbau des Post- und Fernmeldewesens in der französischen Besatzungszone nach 1945". Archiv für deutsche Postgeschichte (1979)1, p. 43.
60 Reinhard Schulz, Geschichte mit Zukunft. 30 Jahre Fortschritt in der Nachrichtentechnik. 1945 - 1975, ed. by AEG-Telefunken, Backnang (not dated), p. 9.
61 Erich Müller-Mees, "Der Selbstwählferndienst bei der Deutschen Reichspost", Europäischer Fernsprechdienst (1942)60, pp. 30-31.
62 Emanuel Hettwig, "Planung von Fernsprechnetzen bei den Eisenbahnen", Siemens-Zeitschrift 17(1937)2, pp. 65-71.
63 M. Hebel and R. Winzheimer, "Landesfernwahlprobleme und Vorschläge zu ihrer Lösung", Jahrbuch des elektrischen Fernmeldewesens 7(1953) p. 148.
64 Martin Hebel, Planungsvorschläge zum Wiederaufbau des deutschen Fernsprechnetzes mit Fernwählbetrieb, München: Leibnitz, 1948, pp. 3-4.
65 Hermann Düll, "Der Aufbau des Fernkabelnetzes in West-Deutschland", Jahrbuch des elektrischen Fernmeldewesens (1952), p. 147.

66 Geschäftsbericht der Deutschen Bundespost über das Rechnungsjahr 1952, ed. by Bundesministerium für das Post- und Fernmeldewesen, Frankfurt, [1953], p. 39.
67 W. Pietsch, "Wählsternschalter und Gemeinschaftsanschlüsse in den Fernsprechortsnetzen", Postarchiv 69(1941)2, pp. 143-166.
68 Hans Steinmetz and Dietrich Elias, Geschichte der Deutschen Post, Vol.4, 1945 bis 1978, ed. by Bundesministerium für das Post- und Fernmeldewesen. Bonn, 1979, p. 599.
69 Internationale Fernsprechstatistik. Stand 1. Januar 1984, ed. by Siemens AG. München, 1985, pp. 39,49.
70 W. Clausen, "Der Selbstwählferndienst und seine Auswirkungen auf die Benutzer unter besonderer Berücksichtigung der Erfahrungen im rheinisch-westfälischen Industriegebiet". Jahrbuch des elektrischen Fernmeldewesens 7(1953), pp. 124, 127, 136-138.

CHAPTER 8
THE UNITED STATES AIR TRAFFIC SYSTEM: INCREASING
RELIABILITY IN THE MIDST OF RAPID GROWTH[1]

Todd R. La Porte

1 Introduction

United States air traffic system (USATS) providing both air navigation and traffic separation became a nationwide governmental service in 1936 after two decades of expanding private and public activity. Within fifty years, this system has grown into an extraordinary matrix of 600 airports and 300,000 miles of airways in continuous flux and motion as millions of people and mountains of freight (and air mail) are shepherded throughout the U.S.. It has been a remarkable development of a very large-scale, publicly owned technical system with quite different properties than the other systems discussed in this book. It is at once, more far-flung and complex, and less integrated and dependent upon technologies as a means of coordination. It has a different relationship to the national state. After a brief review of the dimensions of the USATS, we turn to these properties, suggest their importance for more general understanding of large-scale technical systems, and go into more detail in describing the extraordinary development of the USATS.
    The initial stimulus was transporting mail by air. Both early airmail and airways services were managed by the U.S. Post Office Department until 1925, when private contractors took over the mail services. Air mail flights had expanded from the first regional (daylight) links in 1918 between New York and Washington, D.C., to reach across the continent with the development of night-flying navigation aids.[2] Rotating beacon lights set up every 10 miles guided low-flying pilots over 2,000 miles of lighted airways between New York; Dayton, Ohio; Chicago; Cheyenne, Wyoming; to San Francisco (with a spur to Los Angeles). Coast-to-coast runs took 34 hrs. 20 min. westward, and 29 hrs. 15

min. eastward in clear weather, with airplanes travelling at an average air speed of 100 mph. In the first months of service in 1918, 66,555 lbs. (about 33 tons) of mail were flown at an average speed of 72.6 miles per hour. By 1925, there were 96 planes in service. Regular passenger travel also was begun in 1925 in the eastern United States and by 1930 the five major airlines had carried about 400,000 passengers.[3] By 1927, this growing airways system was handed over to the Department of Commerce.

Full use of airplane capabilities awaited the technical developments necessary to deal with blinding weather, the ubiquitous enemy of pilots. By 1929, the accumulated inventions of the artificial horizon, directional (heading) gyro, and improved altimeter in the cockpit and ground based radio navigation ranges combined to provide the instruments necessary to maintain aircraft altitude orientation and navigation information while "flying blind" in dense clouds. Insuring the capability for "all-weather" flying and navigation through increasingly accurate instrumentation and an expanding network of ground based navigation and communication capabilities continues to be a priority in the Air Traffic Control (ATC) system.

Early institutional developments set much of the basic pattern that still persists.[4] Government subsidies of air mail contracts in the late 1920s provided the infant industry a stable market and prompted techniques that became the basis for airline operations. They also laid the foundation for the present Federal role in providing navigable air routes and other air traffic services. With considerable encouragement from the aviation industry, the Federal government reluctantly accepted responsibility for licensing pilots, inspecting aircraft and supervising the use of airfields and navigation safety.

Due in large part to the controversy surrounding the case of General Billy Mitchell and the use of air power for military purposes and the work of the President's Aircraft Board (1925), the military was separated from civil aviation with the establishment of the Aeronautics Branch, (to become the Bureau of Air Commerce in 1934) within the Department of Commerce.[5] In 1940, experimenting with various regulatory and administrative arrangements, President Roosevelt re-organized the Civil Aeronautics Authority. Economic regulatory functions were placed in a new Civil Aeronautics Board. Navigation and airways management functions remained under the Civil Aeronautics Administration.

By 1940, an embryonic operational air traffic management system was nearly in place and its essential, persisting dynamics established. Several communications and navigation aid innovations had been deployed in the mid-1930s. Twelve airway traffic control centers were spread round the country and airport and airway traffic procedures were standardized.

Finally, an important - political - element in air traffic system development emerged in response to the hazards of air flight. The hazards were made very clear in 1935 when Senator Bronson Cutting was killed in a highly publicized crash. Both the obvious benefit - and threat - to individual leaders became vividly evident. This event focused Congressional attention on the Air Traffic Control System and greatly accelerated air navigation modernization programs. One could describe the repeated pattern of Congressional alarm and complacency as a stimulus/response.

The present system is far-flung, the activities within it intense: thousands of aircraft depart and land at peak periods in the mornings and afternoons in the daily ebb and flow of traffic. Annual traffic in 1980 was over *47 million* hours of commercial and private aircraft flight time, *380 million* passenger enplanements, and *200 billion* revenue passenger miles.[6] Two tiers of airways separate the high flying jets from slower propeller driven craft. High altitude airways are used by a mix of civilian and military airplanes travelling at over half the speed of sound (about 6-7 miles a minute). High flying aircraft are guided through their slower, lower and more numerous brethren to airports with runways over a mile and a half long. Any aircraft above 18,000 ft. must be logged-in, visible on an air control radar screen, and in direct radio communication with an air traffic control center.

The air traffic system is based as much on the cooperation of large cadres of pilots, air controllers, and airways facilities providers as on the array of sophisticated electronic, communications and computer technologies they operate. Its overall performance is remarkable: in 1980, U.S. air traffic controllers handled an aircraft across an airspace *73 million times* with no mid-air collisions. (See Table 1: Elements of USATS and Changes in Scale.)

The system's growth has been phenomenal; its record of safety, astonishing. It affords safe passage at any hour, in almost any weather - usually to any airman who is qualified to seek it. It is a system that spans the globe, and reaches to heights where the curve of the earth is visible. What has been the path of its development; the princi-

Table 1: Elements of Air Traffic System and Changes of Scale: 1940-1985

	1940	1980	[1986]
Airports:			
(paved, lighted)	776	5,830	6,720
Aircraft:			
Prop	--	238,160	246,540
Jet	--	5,869	8,174
Air travel (in 1000 hours):			
Domestic Air Carriers	710	6,250	7,360
General Aviation	3,200	41,000	34,063
(Revenue Passenger Miles in Millions)	1,050	200,000	270,100
Air traffic control:			
Airway miles (1,000s)	32k	296k	325k
Nav. aids (all types)	340	2,090	2,261
Landing aids (all types)	--	988	1,166
Facilities (terminal/route)	11	527	525
FAA employees (1,000s)	5k	55k	47k
Aircraft Handled per yr (in 1,000s)	(1945)		
Air Carriers	2,610	23,600	26,373
Air Taxi	--	7,230	11,794
Gen. Aviation	410	36,720	30,523
Military	2,610	5,990	6,328
Total	5,630	73,540	75,020
Safety Record:			
Air Carriers (Dom. Ops.)			
Accid. per 100K hrs	4.2	0.22	0.22
Fatal Accidents	3	0	4
Fatalities	45	0	197
Fatal accid. per 100K hrs	0.42	0	0.05
General Aviation			
Accid. per 100K hrs	108.4	9.2	8.6
Fatal Accidents	232	618	490
Fatalities	359	1,239	937
Fatal accid. per 100K hrs	7.2	1.7	1.53

ples that have informed it? Are there lessons to be learned from its evolution that alert us to the deeper dynamics of large technical systems?

2 Conceptual perspectives

In this chapter, the United States' huge air transportation system is viewed as a complex socio-technical system of moderately linked organizations shaped by the country's political culture. The system's rapid growth has resulted from a mix of public and private interests facilitating financial, operational and technological advances. The outcome is a complex, quasi-formal mix of private interests and firms and several government agencies. It is a large, highly integrated socio-technical system with essentially no competitors.

A full discussion of the entire U.S. air transportation system is well beyond the scope of what is possible here. It would include attention to the technical development of a growing variety of airplanes, airport construction (heavily subsidized by the Federal government), and the role of the U.S. military in the development of the communication and coordination infrastructure. It would attend to the politics and growth of popular non-commercial flying (so-called general aviation), as well as government regulation of aircraft structures and pilot performance and safety.[7]

Each of these components is itself complex and large-scale. Each is linked to important segments of American society: networks of technical elites, operational managers, industrial and governmental organizations and legislative interests. Together, these actors and organizations comprise a public/private sector of critical importance to the economy, national security and social life of the nation.

Our attention centers on "United States Air Traffic System" (USATS). It is a web of technologies and institutional relationships linking the components of the larger U.S. air transportation system through continuous coordination of aircraft. The system's primary institutional embodiment is the Federal Aviation Administration (the FAA) and its predecessor agencies.[8] Secondary notice is taken of the air carriers and other "users" of the system.

The USATS, unlike EUROCONTROL its younger and much smaller

brother in western Europe, is predominantly funded by resources from the general tax fund.[9] Conceptions of economic development do not adequately explain USATS development. Instead, I draw, in part, from developmental concepts as heuristic metaphor, and, in part, from the literature of organizational theory. Our purpose: to understand the development of an organization that manages a growing volume and complex mix of traffic with increasing scope, safety and reliability.

The time frame of this review is limited, beginning with the early days of the system in the 1940s and ending in 1980, just before its third major institutional crisis - the tumultuous strike of the Professional Air Traffic Controllers Organization (PATCO).[10] This strike, its aftermath in operational travail, and the recent problems of the FAA (brought on by a combination of the deregulation of air transport and a controller cadre working continually at or near full capacity) are fascinating in their own right. Understanding this crisis, however, requires a good bit more than the story discussed below.

Parts of this story have been treated in institutional histories of the Federal Aviation Administration (FAA) and its predecessor agencies,[11] in descriptions of the technical systems planned by engineering groups to carry expected loads,[12] and in evaluations of FAA operations.[13] All of this literature speaks to those who already know a good deal about the technical and operational aspects of national air traffic systems. None provide a perspective which can directly assist us in teasing out insights into the development of the air traffic system as a social system. A conceptual frame is needed which brings the technical languages of machines, structures and operations closer to the languages of social science and social history.[14]

An integrating frame

A major step toward integrating technical and social science perspectives can be taken by conceiving of technical systems as social organization. In this view, the technical design and operational imperatives become guides to operator and managerial behavior.[15] From a social science (or public policy) view, unless a technology becomes widely spread (or is likely to become so) it is a trivial activity. Widespread distribution or deployment of a technology necessarily requires some form of large-scale social organization. It may be decentralized as in the manufacture and distribution of personal computers. It may be physically and organizationally widespread and highly integrated like the distribution of electrical energy through large regional, national or even multi-national grids.

In this view, the techno-organization animates or gives social expression to technical possibilities. This perspective challenges us to examine the properties of technical designs and engineering systems in terms of their organizational requirements and imperatives. It leads us to explore the relationship between the designers' views of operational necessities and the implications of implemented designs for the behavior of operators who man the system.

Conceiving of technical systems in this way enables us to use organization theory to understand the social dynamics of techno-organizational systems, and the patterns of adaptation they exhibit in different situations or environments.[16] A techno-organizational system, then, is shaped, internally, by the social requirements and social properties of technical operations inherent in its engineering designs and, externally, by cross-cutting pressures from its "host society."

When we conceive of the USATS in this way and compare it to the other LTSs under discussion in this volume (telephone, railroads, electrical power), important similarities and differences are evident. These are outlined schematically in Table 2. The similarities are reasonably obvious and we merely list them. The differences point to several important dimensions that would be useful for more general comparisons of LTSs.

Functionally, the USATS is a complex "sub-system" of the larger "whole system" of the U.S. air transportation industry. It is a lesser included, crucial element, in air transport operations. It is also much less fully integrated with its system neighbors than the elements of other systems discussed in this book. Put another way, the "hold" over USATS by other sub-components is a good deal less tight than that evident among the components in European or U.S. rail, electrical power, or telephone systems. It is less tightly coupled, physically, technically, and administratively to its system symbionts. USATS has experienced many of the same dynamics in its development from a small regional to a national network, as our European comparison systems, although changes have occurred more quickly in the U.S.. The logic of national benefit and integrated technical scope have been more immediately compelling. At the same time, the aviation technologies of flight and coordination are less integrated with each other than we see in our comparison systems.

Table 2: Similarities and Differences Between Air Traffic Systems and Other Large-Scale Technical Systems

1. Similarities (parallel components and connectors)

   Central Input Facilities (Initiating activity)

Airports	Rail heads and roundhouses
	Power generators
	Phone exchanges

   Network Connectors and Control

Air Traffic Control	Rail beds and traffic control
	Transmission networks and Switching centers
	Phone networks and Exchange/control systems

   Network Users

Users' aircraft (Commercial, General Aviation, Military)	Rail cars
	Electricity
	Telephone messages, Data transmission

2. Differences (ATS vs. Other LTSs)

System level:	
Sub-system	vs. Whole systems
Rate of National Development:	
Relatively very rapid	vs. Sustained regional development
Degree of Technical Integration:	
Relatively disjointed	vs. Compact and tightly coupled
Degree of Personnel Integration:	
Full operator involvement	vs. Operator as machine monitor

Airplanes and pilots can operate with more autonomy than trains, telephone services and electrical power systems. The connective networks are much less dominated by physical objects - rails, wires and power grids.

Finally, an air traffic (sub)system is largely a mental rather than a physical construct. It has no visible, concrete supporting connectors. The system must be "seen in the head," a mental construct recognized by thousands of people, (controllers, pilots, facilities managers) in order for "it" to be operative. U.S. Air Traffic Control (ATC), the operator/controller of the USATS, is the arbiter of the mental maps and procedural agreements guiding the behavior of its members. These are quite detailed, with many critical aspects, and must be known and followed by many, many users in order for the system to work highly effectively and reliably. This aspect is much less evident for telephones, electrical circuits, or railroad systems.

For the comparative objectives of this volume, it is important to keep these characteristics in mind as we describe some of the salient aspects of USATS development.

3 The development of USATS: External and internal guiding dynamics

The USATS has had an almost unbroken path of vigorous expansion. Such a pattern requires, at least, a high degree of agreement on system purposes and functions. Throughout its history, the USATS has been the object of an extraordinarily high degree of consensus about its mission. All of the major actors within and outside of the system have agreed that:

- Flying is intrinsically valued and air travel produces a major social benefit.
- All those who wish (and can afford) to fly should have the technical and operational means to do so.
- Due to increased demand for flight, increased technical capacities for aircraft, airports and coordination of aircraft aloft are required; It is the responsibility of the Federal Government to assist this development.

There has been an underlying political agreement that access to air travel via either private means or commercial carriers is very nearly a public right.[17] (This has only recently been questioned.)

The result of this consensus has been a readiness, if not always an ability, to respond favorably to proposals for increased resources for development. Indeed, during the time of our interest, the U.S. Congress had never reduced the amount of money requested by the FAA in support of their air traffic control function.[18] Favorable treatment depended on the degree to which needs could be established and programs justified on the basis of meeting operational criteria. These criteria set the framework for the logic of development, and shaped the character and intensity of energies propelling organizational growth.

External demands from the host society have been constant, if potentially contradictory. The public (and especially its Congressional leaders) demands a system which:

- Is always safe;
- Carries anyone, anywhere, anytime (and is always safe);
- Enables private carriers to make a reasonable profit (while always being safe);
- Requires only modest coordination expenses of carriers, and the flying public. (Secondarily, keep costs for governmental administration moderate in terms of the level of safety and ease of traffic movement provided.)

From the earliest days of air travel in the U.S., there has been a strong emphasis on reducing the risk of operating an inherently hazardous technology. The economic success of air travel depends, in part, on the public's perception that using the service "can be habit-forming," i.e., one can do it time and again and survive. It is an activity of special utility to busy elites. Some of these elites are U.S. legislators whose political success is predicated on being able both to attend to the nation's business in Washington *and* maintain contacts with home constituencies often many hundreds (sometimes thousands) of miles from the capitol. Many of these legislators take an active interest in the quality of air traffic management, especially as it pertains to the movements in and out of one of the two airports the FAA had managed directly - National Airport across the Potomac River from Capitol Hill. (The other FAA airport - Dulles International Airport - is also in the Washington area.) One of the peculiar properties of the

U.S. air traffic system is the degree to which its performance is visible to those who have a direct influence on its funding and regulation.

The twin pressures from the travelling public and elites for extraordinarily reliable and safe performance resulted in a system - one of several large technical systems in the U.S. - that has attempted to achieve failure-free operations. That is, the goal of failure-free performance is a central objective of everyone in the system. This drive to achieve very high levels of operational reliability and the demonstrated effectiveness in nearly reaching these goals year after year qualifies the system as a "high reliability" organization.[19] It is a quality that has had an overwhelming impact on the character and shape of the system's evolution.

Technical systems, then, are initially shaped by the operating requirements and social properties of technical operations that are inherent in its technical design. In the operation of the air traffic system, these imperatives were (and remain):

- *The Technical/Operational imperatives* to provide accurate, unequivocal information about location and intention of every aircraft; procedures which eliminate or drastically reduce the likelihood of disoriented aircraft or unexpected convergence of aircraft aloft, and assure timely guidance information to aircraft operators so that no aircraft "loses separation" from another or has a near collision or, most especially, a mid-air collision. The operative goal is to avoid "loss of separation," i.e., to allow two aircraft to come closer than 5 miles apart (and 1,000 ft. in vertical separation.) This is an absolute criterion for controller performance. If a controller suffers a moderate loss of separation between two aircraft s/he is working *three times during their whole career,* they are discharged.

- *The Technical/Managerial imperatives* to expand an integrated network throughout the nation and strive for optimum internal activity, interaction, and density of flow. The result was/is efforts to "pack the system," specifically, to press for headway between aircraft just above legal separation limits - now 5 miles at altitude, and 3 miles near airports under visual flight rules (VFR).

This combination of "imperatives" leads to a fundamental and abiding tension between safety and reliability on the one hand, and efficiency, on the other. In operational terms, tensions are between those who directly benefit from perceptions of safe systems - commercial pilots, air traffic controllers, Congress (and passengers) - and those who must pay for it - air carriers, general aviation pilots, and the administering agency, its political/budgetary overseers. Users press for the resources and regulations necessary for totally safe commercial flying conditions; payees worry that the technical and regulatory safety and capacity requirements are more costly and constraining than necessary to keep air traffic moving economically and safely.[20] This has frequently pitted the following pairs against each other.

Invest in the system	vs.	*Avoid overinvestment*
Airline Pilots Associations		Airline management, and General Aviation Groups
Flying Public		Taxpayers Groups
Air Controllers Associations		Agency Management
Congress (and later, the National Transportation Safety Board)		Executive Office, esp. the Office of Management and Budget

This is a rich stew of advocates and watchdogs. It is fruitful ground for conflict over means and has the potential for exploitation. Much of the development story of U.S. Air Traffic System reflects such dynamics.

4 The development of USATS: Growth and consolidation

The USATS' maturation has been characterized by strong technical advocacy, institutional turbulence, extraordinary growth and astonishing reliability. The central developmental dynamics swirl around the need to manage a growing volume of complex air traffic while anticipating and implementing the technical transformations necessary to keep safely ahead of demand for air traffic services. Operational requirements consist of maintaining a cadre of dedicated air controllers and airway facilities employees who give social animation to the technical systems of communications, electronics and procedures. Technical planning and development requirements call for advanced engineering, solutions to demanding (and interesting) technical problems and the deployment of costly new systems likely to change the working conditions of the operator cadre (and alter their relationships with pilots).

Early FAA leadership was in full accord with both Congressional and industrial leaders: increase the use of air transport (rail transport was the implicit comparison).[21] There was a vigorous program of airport construction and improvement, and, in the pre-war late 1930s, a sense of urgency and then action to promote the growth of aviation infrastructure in preparation for hostilities. Early technical developments of air-to-ground communication, low frequency radio ranges and standardization of procedures for flying by instrument flying rules (IFR) had improved the capacity to identify and locate precisely the flight path of an aircraft. Controllers were trained to use coordination procedures and "flight strips", manually enter a paper strip for each aircraft aloft, then track the aircraft across airways, routing it in place in the sequence of other aircraft before and after it. These capacities and procedures improved service and allowed effective coordination among aircraft separated by a minimum of 10 minutes or 10 miles headway separation. The system - in the midst of its first major technological phase - was established and "in equilibrium" just prior to WW II.

The war brought substantial increases in traffic, technical developments and institutional challenges that set the stage for the FAA's first crisis. The character of the first crisis typifies subsequent problems and developmental dynamics. FAA and military responses to national defense requirements resulted in rapid expansion of communication service networks within the U.S., the deployment of FAA personnel to operate airport air traffic control towers to facilitate defense activities, and the establishment of provisional rules of air navigation. Military needs overwhelmed all others and the FAA functioned in large part as a civilian adjunct to military aviation and defense requirements.

During the war, military aviation developed new air navigation and air traffic technologies complementary to those of the civil aviation system. Military systems advanced beyond those employed for civil aviation, especially with the development of radar and its capability of "seeing" aircraft many miles from an airfield. Military commanders became de facto managers over many in the civilian controller cadre.

In 1946, immediately after the war, there was a rash of activity attempting to reorient the management of U.S. air traffic system for peacetime conditions. As the system had grown, it had become dispersed and its management structure ambiguous. It was time to re-assert civil control of air traffic management.

The Department of Commerce was authorized to take over the operation of military air navigation facilities overseas. Scattered administrative and training units were consolidated in Oklahoma City, where all the FAA schools were to be centered. Joint research and development policies were established to assure continued technical development and the application of military technologies to civil air uses. Common civil-military instrument flight rules (IFR) were officially issued. The President established the Air Coordinating Committee by Executive Order with the responsibility for coordinating national aviation policy. The International Civil Aviation Organization (ICAO), the authoritative international standard-setting body, assembled representatives of 60 foreign states for a demonstration of U.S. air navigation and traffic control equipment and techniques at the FAA's Evaluation Center in Indianapolis, Indiana. This move was influential in ICAO's later decision to recommend acceptance of the U.S. systems and techniques as international standards.

These post-war activities reflected a deep and persistent tendency for leaders of air traffic systems to coalesce administratively as well as a tendency to eliminate institutional ambiguities which might be the source of operational uncertainties. They concentrated training operations, agreed on common standards, used institutional mechanisms to coordinate policy. Above all, they attempted to limit the likelihood of uncoordinated competition.

There had been earlier attempts to move in this direction, but pre-war civil aviation had been struggling for initial viability. Before WW II, airways were not crowded; the problems of safety were not yet closely related to the real likelihood of mid-air collisions. However, the rapid growth of aviation activities, the blossoming of military facilities and activities during the war years, and the general reluctance to raise post-war types of administrative matters until the war was over resulted in a general sense that the system could become inchoate and disorganized as demobilization got underway.

For some technical "systems," e.g., the automobile or aircraft production, a "disorganized" sector means freedom to compete, possibly to prosper. Monopoly or finely grained coordination, the intent of the 1946 developments, is not preferred by those who stand to gain from competition. In the case of USATS, we see another tendency: the drive to reduce sources of ambiguity or conflict that might be the root of operational surprise. It is a tendency likely to be shared by all technical systems that have a relatively high level of perceived hazard.[22]

Technical developments also serve to reduce operational surprise. In addition to institutional coalescence, 1946 was the year in which perhaps the single most important technical advance in air traffic control was introduced - the radar equipped control tower for civilian flying. This technology was first installed at Indianapolis Airport. (It was a modification and up-grade of radar developed by the armed forces.)

This development signalled the end of the first major technical phase of U.S. air traffic system development. The predominant coordination technique had been a manual/voice reporting system of "flight strips and shrimp boats" (small cutouts moved about a navigation map tracking the location of an aircraft as reported by the pilot). The manual/voice system would be supplanted either by a combination of radar, improved high-frequency navigation aids (VOR) and instrument landing systems (ILS) to improve pilot control during landings in foul weather or by "ground controlled approach" (GCA) in which the aircraft was "talked in" by operators scanning the plane's location and glide slope on specially designed radar. Radar would greatly improve the capacity of ground personnel to identify and assist aircraft aloft. As importantly for the development of the airways system, the *omnidirectional* VOR capability exploded the number of courses available for navigation constrained in the past by the ubiquitous *four* course radio ranges.

In a sense, the original system had nearly filled up. With the generous margins for error necessary in the manual/voice reporting-based system, peak time air traffic near the most used airports was approaching full capacity. Increased system capacity was required. Radar, the new instrumentation and added radio and telephone communications between control centers provided it. This enabled controllers to increase substantially the number of planes that could be worked safely.

New technologies made the controller task less problematic when handling individual aircraft but more complex when dealing with up to 12 to 15 aircraft simultaneously. It also raised the question of how to deal with the situation if the newer, more sophisticated, more vulnerable technology failed. Would the controllers be blinded? How could they re-establish their picture of where everyone was? As radar was introduced, the original system was not replaced. Rather this non-electronic, "cannot break" system is still manned and exercised, operating in parallel with newer systems, "on call" as a continuously available backup.

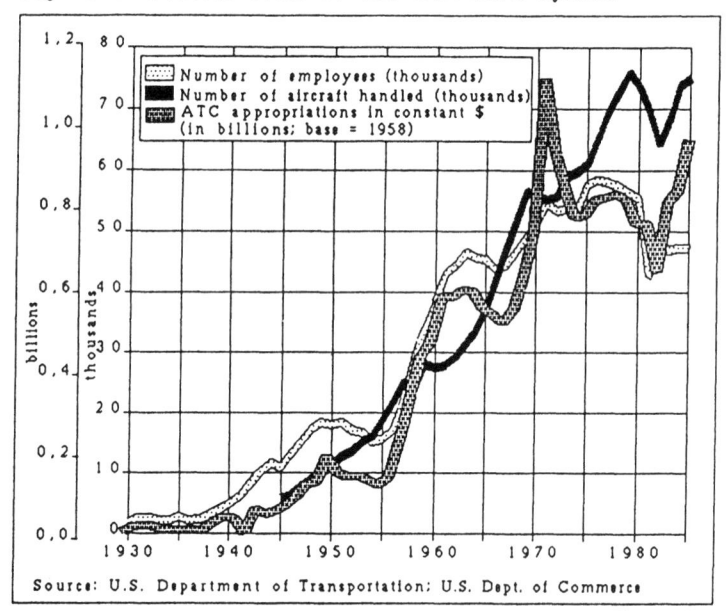

Figure 1: General Scale of the U.S. ATC System

Source: U.S. Department of Transportation; U.S. Dept. of Commerce

Radar gave controllers an independent source of information on the location and disposition of aircraft. The first relatively primitive, sweep radar was augmented by a series of technical changes that systematically reduced the controller's dependence on aircraft captains for flight information. It thereby increased air traffic safety and reduced pilot autonomy.

As radar was deployed to airports and the air route traffic control centers (ARTCC) that monitored the airways between airports, the whole system could handle more aircraft simultaneously. The radar surveillance system was complete: the skies rapidly became more crowded.

One view of FAA's overall developmental pattern is shown in Figure 1. This charts the annual number of employees, and financial resources (adjusted for inflation) available to the agency. A third curve - the growth in actual use of air traffic control services - is laid over the other two. The three curves could be expected to follow parallel paths: however, they are disjunct for three periods and point to times of strain and change. Each of these periods is discussed below. (Figure 2 shows the gross activity load placed on the air traffic control system by different branches of aviation.)

Figure 2: Gross Activity Load on Air Traffic Control System: Combined Aircraft Operations of FAA Towers and IFR Aircraft Handled at Enroute Control Centers.

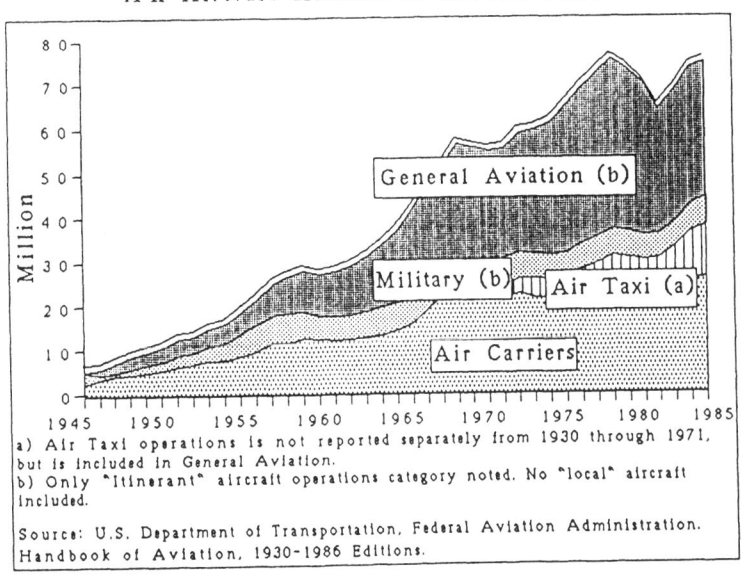

a) Air Taxi operations is not reported separately from 1930 through 1971, but is included in General Aviation.
b) Only "Itinerant" aircraft operations category noted. No "local" aircraft included.

Source: U.S. Department of Transportation, Federal Aviation Administration. Handbook of Aviation, 1930-1986 Editions.

The Korean War produced the first period of strain. Air activities increased over 100 percent from 1950 to 1956, while the FAA's budget and manpower levels declined significantly. The FAA was again part of a war effort and controllers, most of whom had been in WW II, buckled down and kept the system together. It was a time in which technical changes and increased traffic flows would significantly complicate air traffic management tasks. In 1951, the number of air passenger miles first exceeded rail-sleeping car passenger miles, (10.7 to 10.2 million). In 1953, airplane speeds could average over 200 miles per hour. In 1956, the first large jet liners carrying over 100 people were certified. In effect, the stakes involved in commercial aviation had doubled: twice as many people could travel twice as fast and twice as high as in the early days. This was tragically demonstrated high in the western skies in June, 1956.

Two commercial airliners flying in the clear, deviated from their normal route to show their passengers glimpses of the Grand Canyon. They collided, killing 128 people. A Congressional investigation resulted and a series of restrictive measures were imposed to control the movement of aircraft at high altitudes. A continental airspace control service was instituted by the FAA requiring all aircraft in IFR conditions (in clouds) above 24,000 feet to be under positive ATC control. (Submission to this service was optional in clear air.) In 1958, a series of three more tragic airline accidents in the New York/New Jersey area triggered a Presidential investigation and resulted in recommendations for positive air traffic control on the main airways across the U.S. For all aircraft flying between 17,000 and 35,000 feet (this included all jet traffic), IFR rules conducted under prior clearance would apply. Visual Flight Rules (VFR) were rejected in these airways regardless of weather. These changes combined to increase the number of aircraft required to use ATC services and lowered the altitude above which aircraft control was required. The result was a sharp increase in controller work loads, and stimulated a need for more controllers.

At the same time, a battle was brewing between civil and military aviation circles. Research and development on more powerful navigation aids was going apace by both the FAA and the armed forces. In the early 1950s, FAA had begun to deploy a much improved Very High Frequency radio beacon (VOR) that greatly improved the accuracy of determining and following specific directional headings and allowed for a considerably more complex airways system. It also had a distance measuring estimating (DME) capability which gave an indication of the aircraft's distance in miles from the radio beacon. Military development groups were developing a different system, the Tactical Air Communication and Navigation system (TACAN), with similar features, but employed different principles and was more robust for the varied types of operating environments they expected, especially aircraft carrier operations.

In 1947, Congress had directed that future technical developments should strive for a single integrated system. The military insisted that its TACAN be the preferred system on national defense grounds. Plans were to install it at military air bases, and on-board naval aircraft carriers; it was the proposed new navigation aid for the next generation of military aircraft. The FAA was adamant, insisting on its VOR/DME system. Many commercial aircraft were equipped to receive its signals; it was already in operation. Views were fixed and for eight years (1948-56) progress on determining the single system stalled. Two major commissions had been charged with resolving the controversy: two had failed. Finally, at the highest level a compromise was struck: VORTAC was agreed upon. The military would use TACAN, civil pilots would get their directional guidance from VOR but rely on TACAN for the distance measuring component. Efficiency flagged in the face of technical aggressiveness and stubborn operational argument. In effect, redundancy was enhanced despite the best efforts of Congress and the White House.

Problems with the civil-military relations were not limited to technical rivalry. The Air Force and Navy still carried out a number of air traffic control functions. On the grounds of maintaining capacity for use in wartime, they wished to keep them. Some way of coordinating and rationalizing the use of facilities and integrating military and civil air traffic functions was needed so they would be compatible with national defense needs.

To work out what was proving a very difficult process, President Eisenhower felt he needed a man skilled in both aviation and the military. His aviation adviser, Air Force General Elwood Quesada, appeared the ideal person. But Congress had provided in the Federal Aviation Act that no career military man, including those retired, could hold the office of Administrator. The General, wealthy enough to retire early from the Air Force, was persuaded to forego retirement benefits and to accept a special Congressional measure allowing him then to become the first Administrator for the new Federal Aviation

Agency. (When he stepped down, Congress restored his military benefits.) Quesada had the job of consolidating civil aviation services and conducting Project Friendship, i.e., to negotiate what military facilities, practices and operations would be transferred to the FAA. This was completed in two years with the transfer of over 2,000 military air traffic control facilities in over 300 global locations to the FAA. The lineaments of the present system were in place.

By the early 1960s, the jet age was maturing. A number of jet aircraft had been certified that carried well over 100 persons. Jet speeds were increasing. Air passenger transport had forged well ahead of both the railways (for long haul domestic travel) and ships (for the Atlantic crossing). In addition to much higher aircraft speeds and flying altitudes, further technical and system enhancements were made.

In 1958 and 1959, the FAA had instituted Continental Control Areas (above 24,000 ft.) and several Positive Control Routes (between flight levels of 17,000 and 35,000 ft.) in which aircraft were mandated to be under Instrument Flight Rules (IFR), have operative radar and radio communications, and place themselves under ATC direction. By 1961, this system was replaced by a national system providing routing direction and radar advisories along three tiers of airways: lower level from 1,200 to 14,500 ft., intermediate airways from 14,500 to 24,000 ft. and high altitude jet ways above 24,000 ft. At about the same time, computers were beginning to be used for aircraft accounting tasks.

This three tiered airways system enabled the FAA to continue serving a rapidly growing aviation industry within a traffic system which had become increasingly dense and tightly coupled. It also further complicated air traffic control operations, and required a parallel division of labor within ATC centers. Total FAA employment had increased to about 30,000. The air traffic system had become a full fledged bureaucracy of sizable proportion.

The FAA established Associate Administrators for Administration, Programs and Development. The airspace system programs included the Air Traffic Service, Flight Standards Service, Systems Maintenance Service and Airports Service providing guidance to seven U.S. FAA regions. Then, partly in response to the increased coordination needs, the FAA reversed a 15 year policy and gave the Washington office direct supervision over programs in the field. This immediately preceded several years of increasing operational and administrative complexity when air route traffic control centers (ARTCC) were being up-graded technically and, as a consequence of better radar and communication capabilities, were consolidated into fewer, more widely ranging ARTCCs.

The mid-1960s brought the second, now operational, period of strain. Between 1963 and 1967, there was some 65 percent overall growth in the amount of ATC traffic. Resources, however, did not follow the same pattern. The FAA's resources and manning levels *dipped* some 10 percent. The U.S. had become embroiled in Vietnam and war costs were soaring. President Johnson was attempting policies of both "guns and butter" and many Federal agencies faced increasing demands for services with stable budgets. While personnel resource levels hit a plateau, work loads increased steadily, due about equally to growth in both commercial and general aviation users.

At the same time, technical, procedural and administrative changes were "rationalizing the system." By 1964, three tiered airways gave way to the present two, and DME (distance measuring equipment) was mandated for all civil aircraft flying above 24,000 ft. Solid-state, real-time computers were introduced throughout the system. As a result, ATC operations were modestly simplified. Advanced radar systems increased the accuracy of aircraft position images. Computer-generated displays of aircraft identification and position enabled controllers to increase the number of aircraft they could handle simultaneously from 12-15 to 20-25. By 1965, the Continental (positive) Control Areas were expanded to cover the whole U.S. These technical and procedural improvements increased individual controller effectiveness.

Communications and administration were also improved. By 1968, the FAA had put in place a nationwide telephone and telex system connecting the central office with the most active airports, ATC regional offices and area managers. Daily conference telephone calls became a standard feature of management coordination. In addition, the FAA and Air Force increased their coordination and eliminated overlaps. ATC played a larger role in defense interception work and Continental Defense Command activities. Key FAA programs were centralized under the Administrator, while several critical functions that varied from region to region, such as the designation of controlled air space in terminal areas, were de-centralized.

Demand, however, grew faster than the system's capacity to handle the volume with ease. The ATC system geared up to handle increased demand. It extended the amount of controlled airspace and improved more airports to enable them to receive ATC coordinated aircraft. Yet budgets and manpower allocations remained relatively constant. The few modest increases were used for capital and computer purchasing programs. The system became more densely packed, the margins for error declined, and working conditions worsened.

This situation drove controllers to consider organizing to secure relief from increasingly demanding, fatiguing and harrowing work conditions. FAA management was unsympathetic. The controller cadres were expected to perform in the face of adversity. They were then and still are part of a "can-do" organization. In many respects, they had a number of the characteristics of a quasi-military management culture. And they endured these conditions for some five years after the onset of the "stable state." In 1968, after considerable internal debate, the Professional Air Traffic Controllers Organization (PATCO) was formed with a membership of 5,000 in the first year. (It was to grow to over 15,000 by 1980.)

In a direct sense, the union that was to attain such notoriety in the early 1980s was yet another fractious product of the Vietnam War. It arose in a context of an increasing number of personnel related issues. The system became more vulnerable to personnel recruiting and retention problems. It also revealed the deep tension between controllers and management that continues to this day. This abiding tension is rooted in differences in judgment regarding what it takes to keep air traffic moving economically and safely. It results in recurrent labor troubles, as well as controversies about the character of technical solutions for future ATC problems.

Shortly after the formation of PATCO, the ATC system experienced its first instance of extreme airport congestion when the New York area airports had a day in which almost 2,000 aircraft were significantly delayed in taking off or landing. For the first time, the FAA was put into a position of having to restrict the use of certain airports. This was the initial break in the FAA's long standing public policy of serving any pilot who sought assistance at the time he/she requested it. The agency was edging into a position of having to ration its service - a process it still has a difficult time carrying out.

During the 1960s, the goals of service to all in a climate of extraordinary safety led to a series of incremental improvements in new technical systems, changes in procedures and air use restrictions, and operating rules that brought considerably more air space under direct FAA control, e.g., through lowering of Positive Control Area altitudes from 24,000 to 18,000 feet, and raised the spector of perhaps having to assign priorities to different classes of aviation. This, in turn, raised the question of the optimum relation between serving commercial, highly professional air crews and companies contrasted with the much more numerous, generally less well trained and equipped, though increasingly well organized association of general aviators.

There was and is the general recognition that safety problems arose primarily from pilots who were less skilled and/or were not under direct control of the ATC aloft. This was the source of the unidentified, surprise aircraft suddenly appearing on the radar scope or inadvertently entering restricted airspace and tangling with a commercial carrier. These were almost inevitably General Aviation pilots, i.e., private and business employed pilots flying unscheduled, irregular flights. (See Table 3 for the comparative safety records of commercial vs. general aviation.)

There has been a steady trend - continuing to the present - toward expanding the positive airspace under mandatory ATC control and increasing the instrument flying skills and navigation equipment requirements, e.g., multiple radios and radar transponders, in order to obtain ATC services. In the interest of overall efficiency and safety, users of the system have been required to increase their skill levels, technical and equipment capabilities and procedural and operational complexity. These changes have squeezed out the General Aviation aviator who has neither the time nor the money to keep highly skilled and to purchase and maintain costly on-board electronic equipment necessary to qualify for ATC service.

The benefits, stakes and costs of reliable, effective air transport were steadily growing. Thus far, however, sharp trade-offs in service had not been necessary. Vigorous activity was continually required to stay ahead of the demands of increased traffic. Higher skills, more information and tighter coordination processes were also necessary to handle increased system complexity and density. Computer-based data links and inflight following and up-dating of aircraft progress, were improved. And more finely integrated landing and navigation systems were introduced.

In the late 1960s and early 1970s, the FAA paid greater attention to the improvement of ATC controller training and retention. The agency expanded its national ATC training facility. Measures were taken to

Table 3: Accident Trends: U.S. Air Carriers (Domestic Operations) and General Aviation, 1930-1985

| | U. S. Air Carriers | | | General Aviation | |
| | Accidents per 100,000 hrs. fln. | | Fatalities/ 100 million | Accidents per 100,000 hrs. fln. | |
Year	Total	Fatal	pass. miles	Total	Fatal
1930	29.43	3.01	28.2	---	---
1940	4.22	0.422	3.0	108.4	7.2
1950	1.90	0.195	1.1	46.6	5.1
1960	1.78	0.286	0.9	36.5	3.3
1970	0.539	0.017	(0)	18.1	2.5
1980	0.22	0.00	0.001	9.9	1.7
1981	0.38	0.06	0.001	9.5	1.78
1982	0.23	0.05	0.001	10.1	1.84
1983	0.32	0.06	0.001	9.9	1.78
1984	0.19	0.01	0.001	9.6	1.73
1985	0.22	0.05	0.001	8.6	1.53

Sources: U.S. Dept. of Transportation, Federal Aviation Administration, FAA Statistical Handbook of Aviation, 1930-1986 Editions.

improve controller work situations. These changes came at a time when PATCO first tested its strength by initiating a three day, small scale, relatively ineffective work stoppage or "sick-out" in June, 1969. The "sick-out" was followed by the organization's first formally called strike in mid-1970. Some 3,000 (of some 16,000) controllers, mostly at the key ARTCCs, walked out for nearly three weeks. Airline schedules were severely disrupted. The issue, as in the earlier "sick-out," had to do with working conditions, pay and benefits. Having made its point, PATCO called off the stoppage during the court ordered show-cause hearing.

Another technical/systems development advance, Central Flow Control (CFC), was quietly introduced at FAA headquarters in 1970. CFC has been critical to the increased coordination of the sprawling ATC system.

This facility took over some of the responsibilities of controlling the flow of traffic from the 21 ARTCC centers throughout the U.S. Linked by telephone and teletypewriters, the facility was able to determine the overall capabilities of the system on a daily basis and issue instructions for restricted air traffic flows into areas that fell below expected capacity. (CFC became immensely important in the FAA's response to

Table 4: Hours Flown by General Aviation and Scheduled Domestic Air Carriers, and Passenger Miles Flown by Scheduled Domestic Air Carriers, 1930-1985

	Hours Flown (1,000s)		Revenue Passenger Miles (1,000,000s)
Year	General Aviation	Scheduled Air Carriers*	Scheduled Air Carriers
1930	--	299	85
1940	3,200	710	1,052
1950	9,650	2,055	8,007
1960	13,121	3,530	30,567
1970	26,030	5,770	104,156
1980	41,016	6,247	200,829
1981	40,704	6,080	198,715
1982	36,457	5,962	210,149
1983	35,249	6,175	226,909
1984	36,119	6,971	243,692
1985	34,063	7,364	270,061

* Prior to 1971, Hours Flown was calculated by dividing the number of revenue miles by average speed per year.
Sources: U.S. Department of Transportation, Federal Aviation Administration, FAA Statistical Handbook of Aviation, 1930-1986 Editions; U.S. Department of Commerce, Bureau of the Census, Historical Statistics of the United States, Colonial Times to 1970; U.S. Department of Commerce, Bureau of the Census, Statistical Abstract of the United States, 1978 and 1987 Editions.

the near national emergency precipitated by the firing of 11,400 PATCO controllers in 1981.)

The third period of strain occurred in the latter half of the 1970s. General Aviation levels exploded. While commercial carriers were more or less constant in their hours of flight time (see Table 4), jumbo jets were introduced. The passenger carrying capacity for commercial flights almost doubled, up to 200-250 per flight and flying speeds rose dramatically. Once again, the stakes involved with safe flight escalated.

The system approached another period of expected saturation. Brisk planning went on in anticipation of changes in the 1980s. A National Airspace Plan was devised which was intended to provide the radar and computer technologies to "tighten the system" even more, packing more aircraft into the airspace, with more finely coordinated traffic control in metropolitan areas hosting an increasing number of airports with enhanced landing capacities.

Air traffic levels continued to increase dramatically. During the same period, FAA financial resources declined in constant dollars. Personnel levels declined as well. The stage was being set for a conflict between controllers and management. This time a robust union was in place.

5 Conclusions. Properties of networked technical systems

From this review of USATS development, what can be learned about the developments of large-scale technical systems? Does this story point to similarities among the systems discussed in the book? I think it does. They are the properties of *networked* LTSs. These are the systems whose benefits depend on the qualities of networks of dispersed facilities and connectors that are relatively tightly coupled. These properties appear to intensify over time - as a function of the scale and complexity of the system.

Networked large technical systems are:

- Tightly coupled technically, with complex "imperative" organization and management prompted by operating requirements designed into the system, i.e., unless operations are conducted in x,y ways, there

are no benefits, maybe great harm can be imagined. (This is a kind of soft technical determinism: either do it my way or it won't work and do good things for you.)
- Prone to the operational temptations of network systems, i.e., drive to achieve maximum coverage of infrastructure, and maximum internal activity or traffic within the network.
- Non-substitutable services to the public, i.e. there are few competing networks delivering the same service. (The more effective the existing systems, the more likely its monopoly.)
- The objects of public anxiety about the possible widespread loss of capacity and interrupted service. (The more effective it is, the more likely the anxiety.)
- The source of alarm about the consequences of failures to users and outsiders of serious operating failures, e.g., mid-air collisions, nuclear power station disruptions, etc., and subsequent public expressions of fear and demands for assurances of reliable operations.

Notes

1 Work for this chapter was supported in part by the National Science Foundation Grant No. SES-8708046, and the Institutes of Governmental Studies and Transportation Studies, University of California, Berkeley. My thanks to Nick Komons, Historian, Federal Aviation Administration, for his careful review and Paula Consolini for able assistance in data collection and review. Remaining errors are, of course, my responsibility.
2 Night mail flights were initiated in 1921 after this was proved feasible by following bonfires provided by volunteer farmers. See W. Leary, Jr., Airmail Pioneers. Washington/D.C.: Smithsonian Press, 1973.
3 C.V. Glines, "ATC's First Half Century", Air Line Pilot (June 1986), pp. 18-23.
4 Cf. A. Stinchcombe, "Social Structure and Organizations" in J. March (ed.), Handbook of Organizations. Chicago: Rand McNally, 1965, pp. 142-193.
5 For much of the historical detail we draw on the very useful chronology of technical and institutional developments in the U.S. Air Traffic Systems. A.E. Briddon et al., FAA Historical Fact Book, A Chronology, 1926-1971. Department of Transportation, Federal Aviation Administration. Washington/D.C.: Government Printing Office, 1974.
6 The statistics used herein are found in U.S. Department of Transportation, Federal Aviation Administration, Statistical Handbook of Aviation, Statistical Yearbooks. Washington/D.C.: Government Printing Office, 1981, 1986 editions.

7 See D.E. Charlwood, Take-off to Touchdown: The Story of Air Traffic Control. Sidney: Angus and Robertson, 1967; G. Gilbert, Air Traffic Control: The Uncrowded Sky. Washington/D.C.: Smithsonian Institute Press, 1973.
8 Note: The Federal Aviation Administration (and its predecessor agencies) have carried on three functions since the mid-1930's. Air traffic coordination and extending and maintaining a national system of airways has been its primary function, occupying some 75 to 80 percent of its annual budgets throughout the years. The FAA also has promotional and regulatory functions. The agency encourages communities to develop more effective and higher capacity airports to facilitate commercial air transportation services. It is quite visible politically, the vehicle for the dispersement of millions of dollars each year for airport construction and improvements in states across the country. The FAA has an important regulatory function, as well: assuring the public that the aircraft operating within the U.S. possess a high degree of airworthiness and that the pilots flying them are highly skilled and fit. To provide such assurances, the FAA has had vigorous programs in pilot certification involving, in 1980, some 800,000 currently qualified pilots; and it conducts technically demanding certification of aircraft prior to their introduction into regular use. An important aspect of these latter activities is the investigation of aircraft accidents in support of the National Transportation Safety Board, ferreting out causes as a basis for improving procedures and/or making remedial changes in aircraft design and configuration.
9 We set aside for the time being, a) the tale of the engineering developments of navigation, communication and aircraft tracking technologies that are the electronic nerves of air traffic identification and coordination, and b) the process and conditions that resulted in the assurance of continuous growth in air travel. There is also a small charge added to each airline ticket and deposited in the Airports and Air Ways Trust Fund. Until recently, this fund has been used mainly for the improvement of existing and the development of new airports.

I assume that aircraft operate very reliably and are flown by very skilled pilots. In effect, we hold constant the predictability of the machines and their operators - the prime objects of coordination by those who man the air traffic system.
10 The fourth crisis is upon us: a clear sense that the system is nearing its capacity to deal with existing traffic in a safe and expeditious manner. This is due to the effects of deregulation and a continued growth in general aviation (non-commercial carrier) flying.
11 See N.A. Komons, Bonfires to Beacons: Federal Civil Aviation Policy under the Air Commerce Act, 1926-1938. U.S. Department of Transportation, Federal Aviation Administration. Washington/D.C.: Government Printing Office, 1978. J.R. Wilson, Turbulence Aloft: The Civil Aeronautics Administration Amid Wars and Rumors of Wars, 1937-1953. U.S. Department of Transportation, Federal Aviation Administration. Washington/D.C.: Government Printing Office, 1979. S.I. Rochester, Take-Off at Mid-Century: Federal Civil Aviation Policy in the Eisenhower Years, 1953-1961. U.S. Department of Transportation, Federal Aviation Administration. Washington/D.C.: Government Printing Office, 1976; R. Burkhardt. The Federal Aviation Administration. New York: Praeger, 1967. (Primarily the administrative and legislative politics until 1965.) R.J. Kent, Jr., Safe, Separated and Soaring: A History of Federal Civil Aviation Policy, 1961-1972. Washington/D.C.: U.S. Department of Transportation, Federal Aviation Administration, 1980. E. Preston, Troubled Passage: The Federal Aviation Administration During the Nixon-Ford Term, 1973-

1977. Washington/D.C.: U.S. Department of Transportation, Federal Aviation Administration, 1987.
12 For example, FAA Facilities and Equipment Programs for Safety: Hearings before the Aviation Sub-committee of the House Public Works and Transportation Committee, Washington/D.C., January 1978, and Plans and Developments for Air Traffic Systems: Presentations at Symposium of Guidance and Control Panel, Cambridge/Mass., May, 1975.
13 For example, Issues and Management Problems in Developing an Improved Air Traffic Control System, Report to Congress. Washington/D.C.:General Accounting Office, 1976.
14 T.R. La Porte, "Beyond Machine and Structure: A Basis for the Political Criticism of Technology", Soundings (Fall 1974).
15 See T.R. La Porte, "Technology as Social Organization: Challenges for Policy Analysis," Working Paper, 84-1. Studies in Public Organization, Institute of Governmental Studies, University of California, Berkeley, January 1984, for an elaboration of this conception and the data requirements it stimulates.
16 This approach parallels C. Perrow's recent work. See Normal Accidents. New York: Basic Books, 1986, esp. Chap. 3. It differs in stressing the variations in the social properties of the technologies. Perrow orders the technologies he examines in terms of their physical properties, e.g., their internal equipment complexity and the degree to which they are mechanically tightly or loosely coupled. Cf. L. Winner, Autonomous Technology. Cambridge/Mass.: MIT Press, 1974.
17 "Any citizen ... [has] a public right of freedom of transit through navigable airspace." Sec. 104, Federal Aviation Act of 1958. In organization theoretic terms, this is a situation in which there is high agreement on goals, and the problem is to gain agreement on means, a much less difficult matter than if specifying organizational goals were also a matter of contest. See J.D. Thompson, Organizations In Action. New York.: McGraw-Hill, 1967, among a number of others, for the decision-making and structural implication of this situation.
18 The FAA has experienced periods of scarce resources (see Figure 1), but these have been visited on the agency from the Executive Office of President. Whatever the agency could get through that office for Air Traffic Control, Congress appropriated. While there have been budget disputes in Congress, and the FAA has not gotten all it sought, these disputes have not concerned the support of the air traffic control function. Rather, they have been about the degree to which funds set aside for "airway and airport development" would be used not only to provide funds to states and communities to build and improve airports but to carry the increasingly heavy costs of operating the system. Until 1980, almost none of these funds, collected as a small tax on each airline passenger ticket, had been allocated for ATC operating costs. These funds are politically precious resources to be dispensed by Congressmen to favored local constituencies.
19 See T.R. La Porte, "High Reliability Organizations: The Dimensions of the Research Challenge." Working Papers on Public Organizations, Institute of Governmental Studies, University of California, Berkeley, March 1987; T.R. La Porte, "High Reliability Organization Project Overview," Institute of Governmental Studies, University of California, Berkeley, July 1987, for an overview of a major project investigating the characteristics of three such large techno-organizational systems.
20 In the public and media mind, the FAA's function as manager of the air traffic system is often confused with its responsibilities for assuring the airworthiness of airplanes themselves. On occasion, airline problems with the physical integrity

of airframe and engines and the FAA's visibility in regulating the manufacture and maintenance of aircraft is blurred in the media. The regulatory role is much less directive than the managerial one and occupies only about twenty percent (20%) of the agency's resources and personnel. See Note 7 above.

21 The Federal Aviation Administration, hereinafter understood to include the FAA and its predecessor agencies, especially, the Civil Aviation Administration (CAA).

22 Indeed, a case can be made that all organizations attempt as much to reduce or contain sources of uncertainty in their internal and external environments as they seek to maximize economic or operational values. See for example, J.D. Thompson 1967, op. cit., and J. Galbraith, Organizational Design. Reading/Mass.: Addison-Wesley, 1977.

CHAPTER 9
THE FRENCH ELECTRICAL POWER SYSTEM:
AN INTER-COUNTRY COMPARISON

Maurice Lévy-Leboyer

1 Electrical power in France - a deviant case?[1]

On the basis of available statistics the French record in the field of electrical power does not seem to match that of other major industrial nations. Production according to official sources did not amount to more than 0.4 billion Kwh at the turn of the century and 1.8 billion Kwh in 1915-19. This is less than 4% of the 30 billion Kwh produced over the same period by Germany, the U.K. and the U.S., taken together. Of course, statistics were quite defective in those early years. They were improved only in 1923 when the French census included for the first time all productive facilities, instead of a sample restricted to the larger electrical plants as in earlier periods. Nevertheless, the gap was a real and persistent one, as indicated by the fact that French output of electrical power only improved (by 5-year averages) from 7.5 to 20 billion Kwh in the interwar years, while that of the three major countries went up from 90 to 230 billion Kwh, leaving the ratio unchanged at 8.4%. In a way it might not seem fair to compare economies so different in terms of population size. But on a per capita basis, output remained at 325 Kwh in 1920-24 and 480 Kwh in 1935-39, i.e. in both cases at only 53% of the three major countries. And the ratios are similar when total energy consumption or domestic consumption (i.e. the use of electric appliances, including light, by households) is used as an indicator.

Obviously one might point out that France was not endowed with natural resources: coal was lacking and distance made power transportation from the Alps or other mountains a costly undertaking at that stage of technological development. However, when one takes into account the engineering tradition of the country, its financial resources, the need to substitute domestic for foreign energy (the bill on that score represented in 1937 15% of total imports) one cannot help thinking that French achievements were not what could have been expected. So, in order to get a better understanding of the whole process and perhaps to find an explanation for the discrepancy, it seems useful to make a comparative analysis of the energy systems as they developed in France and in the three major industrial countries Britain, Germany, and the U.S., using Thomas Hughes's *Networks of Power* (1983), as a basis for comparison.

Table 1: Per Capita Consumption of Electricity in 1938. Ratio between France and Foreign Countries

	France Kwh p.c.	%	Foreign Countries Kwh p.c.	%	Ratio France/Foreign
Traction	32	8	37	5	0.86
Industry	314	74	544	70	0.58
Domestic	77	18	197	25	0.39
Total	423	100	778	100	0.54

Note: The sample includes Belgium, Canada, Germany, Holland, Italy, Japan, Norway, Spain, Sweden, Switzerland, U.K., U. S.

2 The baseline for comparison

In the three countries analyzed by Hughes, one key element that contributed to the success of electrical power was the absence of economic constraints or setbacks in the whole period under review, i.e. 1880-1930. Industries benefited during those years from the quasi linear expansion of the market. Building up power facilities ahead of demand proved successful because of the development, right from the start, of life at night, and later of a more general use of electric light (the peak load in consumption moved from 7-9 a.m. to 6-8 p.m. in the 1920s). Also, because of a continuous improvement in load sharing and scale economies and the consequent fall in the price of energy to the consumer. Price elasticity of demand in large urban centers was high, and consumption was enhanced quite often by the discriminatory pricing system that municipal authorities imposed on public utilities so that they would diversify their markets; in Berlin for instance, they moved from public light (in 1894 theatres, cafés, shops, banks etc. made up 90% of the demand) to traction by 1900, and then to industry. After the war, a fourth market opened up when urban domestic consumption became more common, the percentage of households using electricity, again in Berlin, having increased from 25% to 76% between 1925 and 1933.

Given such a favorable market environment, technical constraints were free to shape the industry through a process that can be described in three stages.

1. In the period 1880-1890, when productive facilities were limited to small central stations, i.e. separate integrated systems that generated and transmitted power for specific uses, the pioneer-entrepreneurs who were still perfecting, promoting, manufacturing, and servicing their own equipment remained in the lead. Edison, for instance, apart from its subsidiaries in London, Paris and Berlin, had already set up 700 central stations in the U.S. by 1886, i.e. after six years of operation, and was producing some 80% of the country's electric bulb output. Emil Rathenau, the German licensee from 1883-84 onwards, also supplied some 250 central stations within fifteen years and a fair share of the German output of electrical appliances. At the same time, however, the industry was not closed to newcomers. A continuous flow of technical innovations were tested and put into practice, such as alternating current unit stations set up among others by Thomson-Houston (who merged with Edison and formed General Electric in 1892), high-voltage transmission of electric power over long distances (the first major lines were realized by Ch. Brown for the Lauffen-Francfort line and by G. Westinghouse between Niagara Falls and Chicago in 1889-93), etc. The power industry was thus centrally organized in that initial stage under a small group of still innovative manufacturing firms operating both in America and in Europe.

2. In the 1900s-1920s the lead was taken by market builders such as Samuel Insull in Chicago. They bought out competitors, forced independent companies to link their street car systems to urban power networks, and took advantage of surplus capacities developed during

World War I to accommodate cheaply the needs of households. In this way large utility companies emerged whose success was facilitated by three technical developments: (1) the installation of super-generators (of some 3-5,000 Kw initially); (2) the diversification of energy sources and the growth in plant size during the war (some German brown coal plants built for chemical production reached a capacity of 250-300,000 Kw); and (3) the general use of water turbines and high voltage transportation systems. The latter had already been tested in California and the Great Lakes around 1895-1901, but were more generally utilized in the 1910s and 1920s at Muscle Shoals, Cossowingo, etc. Increasing scale in output and bulk sales had become a major factor in the development of the system.

3. Over the years, finance was becoming of paramount importance because of the massive demand for energy and the necessity to develop integrated regional or national networks in order to cope with the load factor. This meant the building of large generators that were able to respond to peak demands of short duration but ran the risk of being under-utilized (at a high cost because of the amount of the fixed capital invested) in slack time. A one-billion-dollar outlay to realize a system of that kind had been planned as early as 1921 by W. Murray in America, and $1.3 billion was actually spent on the British grid in 1926-32. But these sums were beyond the reach of any single firm in the field, $50-60 million being the outside limit of the largest banks and public utilities in Germany for instance. Holding companies were to solve the problem. Many had been used by A.E.G. and Siemens from the 1890s onwards to raise the capital they needed for their foreign activities; a United Electric Securities Co. had also been set up by Thomson-Houston as early as 1890 to finance the company's minor customers. After the turn of the century, independent holding companies were to develop, with great success, under the leadership of consulting engineers (such as Charles Stone and Edwin Webster) and of financiers: the 16 largest holding companies had under their control 75% of the American power distribution sector in 1932. Their usefulness in managing, rationalizing and financing the industry is obvious. But they collapsed in 1932 and were then strongly attacked, firstly by local groups of vested interests who had their plants closed or their returns lowered when local utilities were integrated into larger and more efficient systems, and secondly by shareholders who felt they had been abused during the stockmarket boom and fought for

reform. It is this negative aspect that tends to be remembered and detracts from the perception of the true contribution which financial holdings made to the third stage of development. It is therefore important to emphasize, following Thomas Hughes, the equal importance of the three major advances in electricity development, namely central stations, networks of power, and financial holdings.

3   Power development in France: The setting

This development, however, could not be duplicated in the case of France for at least two reasons. First, market demand was not adequate. In the 1880s, when electricity had passed the experimental stage, the economy entered a period of severe depression that left investment, rural migrations and urban concentration at unsatisfactory levels; by 1901, only one fifth of the French population (24% in 1921) lived in towns of more than 20,000 inhabitants. Furthermore, not all local authorities were ready to cooperate with promoters and to develop new technologies for fear of giving them undue privileges through monopoly concession and of causing harm to older and respectable interests such as those of gas companies. It is very much because of the negative attitude of some of them that by 1887, when several hundred central stations had already been built in the U.S., there were only seven central stations in the whole of France and none in Paris; the only venture worth mentioning in the capital, that of the Société Générale d'Electricité, had to close after four years of unfortunate experiences due to the municipality's insistence on lower prices and its refusal to grant a lease. In 1889, on the eve of an International Exhibition in Paris, new concessions were granted but to six separate companies, and again with strict provisions regarding their maximum profit rate, the equipment they had to hand over without indemnities at the end of their 18-year lease, etc. This gave a new competitive edge to the gas companies which in 1895 supplied some 70% of the light used in the capital at one third of the price of electricity. As a consequence, its consumption in Paris increased only from 30 to 175 million Kwh between 1900 and 1913 which is only 40-60% of consumption in Berlin, although population figures in the two cities had an inverted ratio.

Second, in the country at large, electricity production had developed through hundreds of very small plants: there were probably 1400 of them in 1907, 2000 on the eve of the war. The majority among them were hydro units having on average only 250-300 Kw capacity (a few larger ones, of 5-11,000 Kw, were built in the late 1890s to supply power to Lyon, Grenoble and the large southern cities). This gave a decentralized structure to the industry, a feature which was further strengthened during the war since the great majority of military contracts and state loans were granted to these firms in order to develop industry away from the battlefields. This helped to increase the capacity of power production in the Alps and other mountains, but not the average size of the plants of which only six had a capacity of 30-40,000 Kw. Of course large generators five times that size were built in Paris immediately after the war, but this did not restore the balance between thermo and hydro-electrical power. Inflation stopped new undertakings between 1924 and 1926, their long-term profitability being jeopardized by the rise in interest rates and the cost of imports, notably of copper and coal. While costs multiplied by twelve since 1913, the price of electrical power, which was fixed by decrees, increased only by a factor of six. In this way inflation and price control slowed down innovations and contributed to the survival of too great a number of small hydro-plants.

Given such circumstances, the industry's structures and markets could not be equal to that of other countries. Time had been too short for France to catch up, so that per capita consumption and more specifically domestic consumption, kept lagging behind. It should also be recalled, however, that the French economy made a strong recovery in the early part of the 20th century; the growth rate in the production of electrical power kept steady at 11% per annum, equivalent to a doubling every six years over the 1900-1930 period, and it was still at 3% p.a. during the 1930s, despite the general stagnation. Furthermore, a detailed analysis indicates that satisfactory results had been achieved by 1937 in at least three areas. (1) In traction, i.e. in railroads and urban transportation where, already by 1912, the Parisian system was equal in efficiency to those of other big cities. (2) Major steps forward were also made in industry; even if average per capita consumption was still low at a national level (at 314 Kwh, vs. 544 Kwh abroad), it reached a level of 600 Kwh per capita in the North, the East and the South-East, i.e. in the main industrial regions. (3) Domestic consumption had also improved in Paris; its share as a percentage of the local supply was only 16% in 1913, but increased to 28% in 1938, and to 45% in 1944. The unsatisfactory results for the country as a whole were due to the low consumption levels in the rural areas of the Center and the West, where the price of electrical power remained high even though it was heavily subsidized.

Table 2: Regional Patterns in the Consumption of Electricity in France in 1937

	Population %	Consumption Total Billion Kwh	Cons. Shares Domestic %	Others %	Consumption Domestic Kwh per capita	Other	Total
Paris	16.2	2.64	4.5	10.5	116	274	390
North and East	22.9	5.95	2.6	31.1	49	572	621
Southeast	15.5	4.59	3.2	23.5	70	640	710
Southwest and Pyrénées	13.4	2.31	1.6	11.5	51	362	413
Central West	32.0	2.16	2.3	12.2	30	130	160
Total	100.0	17.65	14.2	88.8	57	364	421

Source: Ministère des Travaux publics, Statistiques pour la production et la distribution d'énergie électrique en France, 1939.

In other words, if one takes into account the late start of the second industrialization in France, the material destructions and misallocation of resources caused by the war and inflation, the serious income disparities (regional as well as other) that had always limited market expansion, one might argue in view of the positive developments just mentioned that the country's record was not altogether unsatisfactory, national averages tending to hide some of the brighter aspects of development. The problem therefore is to find how progress was achieved and whether it can be explained by ways that are at variance or in conformity with the sequence presented in Thomas Hughes's account.

4 The formation of industrial groups

In the absence of a central market (until 1907, when the gas monopoly in Paris came to an end), there were no real opportunities for one major company - like Edison, A.E.G. or Siemens - to act as a leader in the industry. The main firm, CCE (Cie Continentale Edison), founded in 1882, quickly lost its status: it merged with its two subsidiaries some four years later and eventually sold out its manufacturing departments in order to act as a simple holding company having interests in various public utilities. In fact, from the very beginning Edison had many competitors, such as Swan, Maxim, Siemens and of course French manufacturers acting independently or under a foreign license, e.g. Bréguet, Sautter-Lemonnier, Gramme, Fives-Lille, l'Alsacienne de Constructions mécaniques, la Compagnie Electro-mécanique (C.E.M.). But, contrary to past experience whereby new industries in France tended to expand in the first stages through a multiplication of firms, a small number of firms soon emerged that were to dominate the field.

In the electrical machinery sector, successful firms were those that had access to foreign patents and capital and specialized in transport. Among them one finds Thomson-Houston, established in 1893 to build street-cars and trolleys; after ten years of experience, the firm had supplied 60% of the French urban transportation network, with strongholds in Bordeaux and Paris. There was also the Empain group, a company founded by a Belgian engineer who had started building local railroads in the North from the 1880s, but specialized after 1894-95 in converting urban transport in Lille and other cities to electrical power; at the turn of the century, the firm eventually took over the Paris subway that had been first started in 1898 by a consortium in which Westinghouse was a major partner. A few other firms, like Grammont and O.T.L. in Lyon, were also active in the field. Being organized as teams of consulting engineers who studied and started new projects, but had so far limited their manufacturing activities and rather imported or sub-contracted the machinery they needed, these firms kept diversifying into new markets, taking up foreign licenses to start new ventures in association with other local firms. Thomson-Houston, for instance, entered in that way the sector of electric meters (in 1891), batteries (1900), railroad signals (1905), bulbs (1912) etc.

This explains that eventually such firms were able to develop and to take the lead of the sector in two major stages. Firstly, in 1902-05, when the construction of big generators for the supply of electric power was undertaken, many firms in order to become autonomous started building or enlarging their own production facilities: C.E.M. opened a factory at le Bourget near Paris in 1902; Empain did so at Jeumont and Longueville in 1904 and 1913, while others like Thomson-Houston absorbed their own subcontractors in 1904-09 and stopped importing from abroad. Secondly, in the pre- and post-war periods, when the electrification of the railways was put on schedule, a series of mergers among major firms took place. At first with limited success, since the inflation of the 1920s prevented long-term investments in railroad electrification. But in 1928-29, a new construction plan was announced and, with it, a final consolidation took place under the lead of firms specializing in heavy machinery, like Alsthom, le Matériel S.W., and Jeumont. In short, although the whole process took place later than abroad and with transport as its main initial sector, manufacturing firms finally assumed the same oligopolistic position in France that one could find at that time in other countries.

In the field of utility companies, the true beginning came after the turn of the century with the rise of firms that had the possibility to operate large plants for captive markets. In that way they were able to dispose of an extra supply of electrical power at low cost and to develop large-scale organizations, the second stage in Thomas Hughes's account. This happened when large generators were built in Paris, first at St. Denis in 1905 by S.E.P. (a member of the Empain group) for the benefit of the subway and bulk-purchasing companies, and later, at Gennevilliers in 1922, by the Union d'Electricité, a company that amalgamated all the firms operating in the suburbs and was to become the largest public utility company in the country. In 1934 these two companies were put in charge of supplying electrical power to the whole Paris region, including the city itself. Under their impulse the size of local production units was increased and a regional network was organized with extra supplies coming from new hydro plants in Central France. This explains that power consumption in Paris at large reached 450 million Kwh in 1926 (a volume equivalent to that of Berlin), almost twice this amount in 1935, and more than one billion Kwh in 1938. At this point, industrial structures had become equivalent to those found abroad.

In the provinces similar systems, i.e. integrated networks of power plants and transport lines, developed on a smaller scale and with the difference that companies had often started at water sites and with small size units, a fact which explains that one of their main concerns was to build transport lines and also additional steam-generators to steady their supply of power during low water seasons. Thus on the eve of the war, Force et Lumière, a company based in Grenoble and Lyon, had six works with a capacity of 45,000 Kw allocated in a ratio of 1 to 9 between thermo and hydro power, while E.E.L.M., a company servicing the Mediterranean market, had 17 plants with a total capacity of 120,000 Kw of which 17% were supplied by 10 thermo units. By comparison with the system in operation in Paris, provincial companies were at a disadvantage in the early 1920s: they had limited urban markets and surplus capacity left over from the war, and therefore problems of under-utilization, of extra cost (due to long distance and loss of power along transport lines), of lack of flexibility, etc. But they devised ways to overcome these difficulties. In regions of surplus production, firms set up cartels with a central office in charge of regulating output, of allocating quotas among members, of opening new outlets and of building transport lines to enlarge markets and thereby to reduce working expenses. One of the very first, UPEPO, founded in the Pyrénées region as early as 1922 by fourteen power firms and the local railroad company, offered an example that was duplicated in many regions and paved the way for a truly integrated transportation system. It was extended both in length and power during the 1920s: in 1930 some 4,700 km of high voltage lines were in operation as against 890 km in 1923. Thus, the building up of power networks was achieved in the 1920s under the lead of two types of firms: 1) the Parisian public utility companies, whose success was based on the technical efficiency of local plants, a diversification of energy sources, and attempts to develop new consumer services, and 2) in the countryside, cartel organizations in charge of diversifying and of broadening markets.

Throughout the period, however, financial resources were a key problem, and this to a greater degree than it was the case with the countries in Hughes's sample. Electrical power required investments on a scale that was reminiscent of 19th century railroads; capital invested in the sector has been estimated in France at 20 billion francs (the equivalent of one billion dollars) in 1933, compared to 1.5 billion francs in 1913. To a large extent, the burden of these large commitments fell upon manufacturing companies. For one thing, it had always been the custom in the sale of machinery to give very long-term credit facilities and even to accept - in part or total payment - the shares of the company that had made the purchase; 67% of Thomson-Houston assets in 1900 and still 46% in 1913 consisted of such securities as compared to 4-8% for the firm's production facilities. Secondly, the high capital intensity of the industry and its limited profitability in its early stages made it a necessity to use external finance on a large scale. In 1913, for instance, reinvested profits amounted to only 5.5% of the 257 million francs invested in the Paris subway. Firms had therefore to assume at the same time the functions of financier and of manufacturer, and because of the long time required to develop public utilities, they were exposed to the hazards of market instability and recurrent economic crises, while the securities they had accumulated were not yet marketable.

To solve this problem, at least in the first stages, firms called upon bankers to assume some of the short and long term financial responsibilities. One third of the board members (and often the president) of the main companies came from financial circles, private bankers (such as Emile Mercet at Thomson, Bénard and Jarislowski in the Empain group) working in close cooperation with the representatives of large deposit banks. But from about 1902 onwards, when manufacturing became a major part of these firms' activities, banks could no longer assume the same functions. Part of the financing, therefore, was transferred to holdings i.e. to companies that were founded to hold the surplus of securities which the manufacturing firms were no longer able to finance. Swiss and other foreign holdings were active from the 1890s in the Alpine region; the Empain group had been developed as a federation of international holdings, based in Brussels, but with strong subsidiaries in France; Thomson-Houston also set up in 1909 the Société Centrale pour l'Industrie Electrique so as to reduce its own commitments, etc.

But the real step forward came only in the 1920s. For one thing, profitability then was restored with the readjustment of concession contracts to raise selling prices and so to protect power companies from the impact of inflation. Secondly, the financing of the whole sector became much more decentralized with the return to convertibility in 1926-28 and the possibility of issuing again industrial securities on the financial market. A survey covering four of the largest public utility companies - with a total of 26 generators built at a cost of 3.4 billion francs over the 1904-39 period - shows that they were both able to increase their stock issues and at last to accumulate profits for reinvestment so that they could assume almost one third (1.6 billion) of the financing themselves. These companies also worked in close cooperation with manufacturers, doubling the number of large generators in Paris, building plants of equivalent size in the country, and - together with the old gas companies and the new holdings - developing extensive power transport systems.

Table 3:   French Electrical Power System

	Capacity (1,000 Kw)						Transport Lines (km)	
	All Plants			Large Plants 50mW+				Voltage
Year	Total	Thermo	Hydro	Total	Thermo	Hydro	Medium	High
1913	740	480	260	--	--	--	--	--
1926	5,260	3,700	1,560	(9) 933	(6) 765	(3) 168	4,455	1,940
1936	9,100	5,800	3,300	(39) 3,661	(28) 2,561	(11) 1,100	5,600	5,750

Figures in parentheses refer to the number of plants.

The impression one gathers from this brief survey is that the three-stage sequence described by Thomas Hughes holds also for France. But even if technology imposed the same structures and policies in each country, the pattern of evolution and perhaps the end-results may not be identical. Among the different countries, those that had started early and accumulated the profits that normally accrue to innovators were able to develop and serve expanding markets in a rational way from the start. In France, the stagnation of the 1880s, the First World War, and other difficulties held progress back until the financial markets opened and until the rise in the (real) price of electrical power made it possible for large firms to restructure the whole industry and finally to close the gap separating France from other countries.

5 The state and the profession

But was the gap really closed? Technically, all power systems were alike. French engineers had been eager from the turn of the century to step up the size of the generators they built and to emulate the efficiency of the German and American plants on which their first works were modelled. It may be recalled that in the early 1930s when the third wave of investment was ending, the three major plants in Paris had an average capacity of some 300,000 Kw, a figure that can be compared to the 5,000 Kw registered at St. Denis in 1905, or to the 50,000 Kw of St. Ouen in 1914. Progress in unit size was obvious. In the provinces, water-works also were increased; they averaged 140,000 Kw in 1932-35 at Kembs, Brommat and Marèges. Furthermore, it had been fully realized that the use of high voltage lines gave new possibilities for the transportation of power over long distances and contributed to lowering its cost; 45-50 km had been a limit for low voltage, but in the late 1920s it was economically feasible to reach some 200-250 km, using a line of 110 kv, and more than 450-500 km at 220 kv. So the use of high voltage lines was steadily expanded, the network reaching up to 10,000 km in 1940. At that date, the French network included eight regional systems that had taken shape as early as the 1920s, and two major lines which came into operation to diversify power supply to Paris; out of about 3 billion Kwh that were then consumed in the capital, some 1.2 billion came from the provinces. These were technical achievements that obviously had brought the French system on a par with the most advanced countries.

The financial situation in France, however, could not be similar to that of other countries. Major investments had been postponed during the years of inflation, and the new capacities whose construction had started in 1928-31 came into production at the very time when the recession was slowing down economic activity. Total consumption, which had doubled between 1923 and 1930 to 17 billion Kwh, now dropped in four years by some 10%. Of course, some sectors may have been better protected from the impact of the depression. This was probably the case with thermo electrical power. Plants in that sector were older, cheaper to build and to manage, and their production was more flexible because of the greater importance that variable costs had in total outlays; moreover their location in big cities made it easier for the public utility companies to extend their markets - as they did in the 1930s - at the expense of gas and other sources of energy. This was completely different for hydro-electrical power. Here capital requirements were higher (1.08 million, in francs of 1913, per Kw, against 0.8 million for steam generators) and fixed costs took as much as 80-85% of total expenditure (cf. Table 4). As a consequence, most hydro-companies incurred heavy losses when demand fell off.

Table 4: Cost Breakdown of Electrical Power in Paris in 1936 (Centimes per Kwh)

	Thermo		Hydroelectric Power Cost			
	Cost	%	Product.	Transp.	Total	%
Coal	5.4	37	--	--	--	--
Materials	1.0	7	0.50	--	0.50	4
Wages	1.5	10	1.00	0.50	1.50	10
Financial costs	5.3	36	7.45	2.70	10.15	68
Taxes, insurance	1.5	10	0.80	0.90	1.70	11
Loss in transport	--	--	--	1.05	1.05	7
Total	14.7	100	9.75	5.15	14.90	100

Source: Philippe Lacoste, Le programme d'équipement hydroélectrique et d'innovation en France à la fin des années 1930 (Université de Paris-X, 1985), after an estimate by E. Mercier.

Furthermore, overdue measures of rationalization reinforced the depression impact. A great number of small unit firms, dating back to the pioneering years of the industry had managed to survive the period of inflation. Even if there had been no recession, most of them were due to disappear once they were integrated into the new power networks. Thus, drastic restructuring policies that intensified the crisis were pursued in the 1930s; this was especially the case in the region of Lyon where several heavily indebted firms unable to survive financially were closed or amalgamated, among others, into l'Energie industrielle, a local company that became at that time the second largest in the country. Besides, pressure groups representing small business firms and local political interests were active in Parliament, lobbying for subsidies, tax exemptions, and price reductions. In pursuing their goals these groups were able to take advantage of an improvement in the income situation of the large power companies whose profits increased in 1930-34, when their new hydro-plants entered production. Eventually they managed to bring electricity prices under government control in September 1934, and, first, to have them reduced by 10% in July 1935, and then frozen at the same level up to April 1937, despite the revival of inflation. This brought to an end the period when investments could be financed autonomously out of profit by individual power companies.

Given the state of overproduction, this was of no great consequence in the mid 1930s. But with the recovery in industrial activity, specifically in the electro-technical sector, the utilization of productive capacities for electrical power was pushed up from 76.6% to 88.4% in the period 1935-37. And new plans were prepared by the Chambre syndicale des Forces hydrauliques, one of the manufacturers associations in the sector, not only to assure a two billion Kwh reserve and thus to ask for the construction of new capacities, but above all to bring into operation a new management scheme for the national system as a whole. The general idea was to give precedence to hydro-electrical power, since it was cheaper to produce (at full capacity) and to distribute through the new interconnecting network, and to minimize thermo-electricity, which had become more expensive with the devaluation of the currency and a doubling in the price of imported coal. Construction plans which were discussed at several meetings of the profession in 1937-38 aimed at developing and regulating the flow of production at the waterworks, while confining thermo-electricity to "un rôle d'appoint compensateur". Also, the high voltage transport network was to be extended in order to treat the whole industry as a single system to be operated more efficiently through dispatching centers.

These plans could well have been implemented by the state. By a series of legislative acts, the French state had assumed from 1906, and more decisively after 1919-25, complete control over the development of the transport system, and over the procedures preparatory to building new power generators. It had also financed, through subsidies, loans and annuities, several works that were of general interest to the country: some 5 billion francs (30% of the total outlay) were spent for the electrification of the railroads and of rural districts; and only 415 million francs (less than 2% of total cost) were allocated up to 1933 for the production and distribution of electrical power. In June 1937, a newly appointed public board in charge of supervising the sector - le Conseil supérieur de l'Electricité - approved a series of new works and their financing by the state. This new scheme would have increased the high voltage transport network by 4,000 km and hydro-production by some 3 billion Kwh, two thirds for the super-plants of Genissiat and Laigle already under construction. However, with the stepping up of the rearmament programme and the outbreak of war, an extension of the public sector proved premature. Hence the planned modernization of the power network had to be financed by le Groupement de l'Electricité, a guarantee fund set up in June 1938 by the two main professional associations, la Chambre syndicale des Forces hydrauliques and le Syndicat professionnel des producteurs d'énergie électrique, and to an even greater extent by the individual companies themselves who had started investing again with the recovery in demand. A total of 18 billion francs were issued in 1938-43. This seems large, compared to the 15 billion francs issued by power companies between 1926 and 1937, but this does not mean that achievements were of equal importance in the two periods. In fact, serious shortages of labor and raw materials during the war reduced the programme to one third of the original plan. Major advances were nevertheless achieved - the establishment of a new national system in November 1942, a shift to hydro-production and to larger unit plants, the development of domestic consumption in the southern regions etc. It would be impossible to explain the upward development of the post-war period (total production rose from 22 to 42 billion Kwh between 1943 and 1953), without keeping in mind these new initiatives.

6 Conclusion

On the whole, the difficulties encountered by the French economy over the fifty years under review were probably too numerous to take its experience as a case in point for a study of electrical power systems. Nevertheless, two conclusions may be drawn from this short survey.

One concerns large technical systems. It is true that the substitution of power networks for the first central stations was a major step forward. They made it possible to increase the unit size of generators, the power and density of the transport system, the level of utilization, and hence to cope with the load factor problem. But the difficulties raised by the operation of large systems in unified markets, problems which were discussed first in Paris during the 1920s and then in trade associations all over the country - about compensatory uses of regional production, of thermo and hydro power, of optimum price and output, etc. - suggest that there is room, in Thomas Hughes's sequence, for another stage. This being the one that started in France by the mid-1930s, at the end of the boom, when operations research and market management became as essential as purely technical innovations for the working of a system that had matured with the passing of time. The second point concerns the factors that explain its development. Even though technological constraints might prevail in the end, it is hard to believe that economic forces, in our case specifically capital intensity, did not play a part in the emergence of the system and did not eventually shape its structures. Market expansion was a key factor and probably eased the process of capital accumulation in Hughes's sample of industrial countries. While in the French case, because of a deficiency in market demand, financial intermediaries - banks, holding companies, trade and state institutions - had to intervene and probably did contribute in the end to hastening the process of modernization by making it possible to use new technologies earlier at a national level.

Note

1 The French electrical power industry has been the subject, in its time, of many monographs. Among them, one may quote:
L. Babonneau, L'énergie électrique en France. Paris, Toulouse: Gauthier-Villars, 1948;
F. Marlio, "L'industrie électrique", Revue des Deux Mondes (1931)6, pp. 335-353;
Et. Genissieu, "Influence du tarif sur la consommation, la recette et le bénéfice dans la distribution de l'énergie électrique", Revue générale d'électricité 35(April 1934), pp. 529-549 and 569-583;
J.P. McKay, Tramways and trolleys: The Rise of Urban Mass Transport in Europe. Princeton: Princeton University Press, 1976;

C. Malégarie, L'énergie à Paris. Paris: Béranger, 1947;
M. Boiteux, "La tarification des demandes en pointe: application de la vente au coût marginal", Revue générale d'électricité 58(August 1949), pp. 321-340;
T. Hughes, Networks of Power - Electrification in Western Society 1880-1930. Baltimore: The John Hopkins University Press, 1983.
J.M. Jeanneney and C.A. Colliard, Economie et droit de l'électricité. Paris: Editions Domat Montchrestien, 1950;
More recent and of a more historical character are:
M. Magnien and R. Pinet, "Le transport d'énergie en France", Revue générale d'électricité (September 1971)11, pp. 40-65;
F. Pézerat, Naissance et développement de quelques entreprises françaises de production et de distribution d'électricité (1880-1939): Etude sur les profits. Paris: Hachette, 1972;
M. Lévy-Leboyer, "Histoire de l'entreprise et histoire de l'électricité" in Association d'Histoire de l'Electricité en France (ed.), L'électricité dans l'histoire. Problèmes et méthodes. Paris: Presses Universitaires de France, 1985, pp. 13-24;
Philippe Lacoste, Le programme d'équipement hydroélectrique et d'interconnexion en France à la fin des années trente. Unpublished M.A. thesis, Université de Paris X - Nanterre, 1985;
A. Broder, "La multinationalisation de l'industrie électrique françaises (1880-1931). Causes et practiques d'une dépendance", Annales 39(1984)5, pp. 1020-1043;
H. Morsel, "Panorama de l'histoire de l'électricité en France dans la première moitié du XXe siècle" in F. Cardot (ed.), 1880-1980. Un siècle d'électricité dans le monde, Paris: Presses Universitaires de France, 1987.
As well as articles of a more limited scope by:
A. Beltran, "Les débuts du réseau dans les villes, 1880-1920", Bulletin d'Histoire de l'électricité (1986);
A. Bertrand-Comiteau, "Force et Lumière, 1899-1914", Bulletin d'Histoire de l'électricité (December 1985)6, pp. 121-136;
C. Bouneau, "L'UPEPO, 1922-1946. Genèse et logique de l'interconnexion", Bulletin d'Histoire de l'électricité (December 1985)6, pp. 75-102;
C. Bouneau, "La politique d'électrification des chemins de fer du Midi, 1902-1937", in Association de l'Electricité en France (ed.), La France des Electriciens, 1880-1980. Paris: Presses Universitaires de France, 1986, pp. 185-198;
C. Bouneau, "La consommation d'électricité dans le Sud-Ouest, 1914-1946", in Association de l'Histoire de l'Electricité en France (ed.) L'Electricité et les consommateurs. Paris: Presses Universitaires de France, 1987, pp. 17-38;
H. Morsel, "Evaluation de la formation du capital fixe dans l'hydro-électricité avant les nationalisation en France", Bulletin d'Histoire de l'électricité (June 1984)3, pp. 5-14;
J. Rapp, "Aux origines de la Compagnie générale d'électricité", Bulletin d'Histoire de l'électricité (December 1985)6, pp. 103-120;
P. Stahl, "Les débats politiques sur l'électricité dans les années trente", Bulletin d'Histoire de l'électricité (December 1986)8, pp. 55-88;
But the most comprehensive works are still in the process of being completed and published. They are:
F. Caron, J.C. Colli and M. Lévy-Leboyer (eds.), L'histoire de l'électricité en France, des origines à nos jours. In preparation;
Pierre Lanthier, Les constructions électriques en France, 1880-1940, Stratégie et financement de six groupes industriels. In preparation.

CHAPTER 10
THE DYNAMICS OF SYSTEM DEVELOPMENT IN A
COMPARATIVE PERSPECTIVE: INTERACTIVE VIDEOTEX IN
GERMANY, FRANCE AND BRITAIN

Renate Mayntz and Volker Schneider

1 Interactive videotex as a large technical system

Interactive videotex belongs to that class of technical systems which are spatially extended and serve to transmit or transport given objects (electrical current, water, passengers, freight, information, etc.) through a network of appropriate channels. Though not a fully independent system since it is normally owned/operated by the telephone agencies and uses their networks for transmission, interactive videotex is more than just another service offered within an already existing technical system[1]. The existence of technical and social components which are specific to videotex make it meaningful to consider it as a large (socio-) technical system and to study its development. Interactive videotex is based on a special set of (linked) computers serving as data banks, it needs special terminal equipment (though *one* alternative is the TV screen connected to the telephone plus decoder), and there are special service providers distinct from the national PTT agency.

For a long time there were only three telecommunication networks: the telegraph and the telephone, both dating back to the last century, and the teletypewriter (Telex in Germany) introduced in the 1930s. This repertoire of telecommunication forms changed significantly only with the advent of microelectronics and the subsequent diffusion of computer technology into the telecommunications domain. Within a short period of time several new telecommunication forms emerged, such as facsimile transmission, data and text transmission systems.

Within this family of new telecommunication media videotex has an important place. It is a form of telecommunications in which not only text and data, but also pictures and graphics can be transmitted[2]. In contrast to the other new telecommunications media, videotex had been conceived for mass utilization from the very start. It was understood, as Forester has put it, as "the spearhead of the information technology revolution that will transform the living room TV set into some kind of electronic supermarket"[3]. Since the late 1970s a number of industrially advanced countries have introduced videotex systems: The British Post Office introduced *Prestel* as a public telecommunication service in 1979[4], France followed with her *Teletel* service in 1982[5], and Germany officially introduced *Bildschirmtext* in the autumn of 1983[6]. These national developments are not only closely parallel, they have also taken place practically under our own eyes, which makes the introduction of videotex a particularly promising case for an internationally comparative study of the development of a large technical system. The research upon which we shall draw in this chapter has been carried out by three different national teams working in cooperation, and may be considered a brief and preliminary version of the full comparative study to be published jointly later on[7].

2 The process of videotex introduction

The introduction of videotex is not only a technological innovation process and a systems development process, it is at the same time a policy process. Quite in contrast to the initial phase of electrical power or railroad system development, videotex has been planned as a nationwide system from the very beginning; a possible counterpart to the process of gradually linking up many smaller, local and/or regional power networks into one big system can therefore only be found at the international level, where attempts are presently made to establish links between various national videotex systems. The reason is obvious: in the three countries considered here, the existing state telephone monopoly offered the central government a focal role in the introduction of this new form of telecommunications from the start, and the existing communication network made the plan of a nationwide extension of the new service feasible. Accordingly, state agents and not private entrepreneurs were the dominant actors in this case of systems development. This might well make an important difference both with respect to features of the process, and to its outcome: In a centrally planned, top-down process of systems development, market forces - profit motives as well as manifest consumer needs - could be of less importance at least in the initial phase, whereas political considerations should play a more prominent role. One consequence could be that the phase model developed by Thomas Hughes[8] turns out to be of less general applicability than he - and we - imagined at first.

If we look at the comparative chronology of videotex development presented in Table 1, it is possible to distinguish - in spite of all differences in detail - three phases in all countries: conception and formation of a policy, experimentation and consensus building, system consolidation and adaptation to usage trends.

Phase 1: Videotex is not a radically new technology, but rather a new combination and elaboration of existing technologies. The new medium emerged not as a technical invention, but as the concept of a new technically based public service, and hence within a policy context. Characteristically, therefore, initial technological development took place in a government laboratory, or the technology was taken over from another country as in the German case. This phase ends with a high-level political decision or even, as in Germany, a series of such decisions (to develop *Bildschirmtext* in 1976, to introduce it nationwide in 1981).

Phase 2: Experimentation and consensus building are parallel processes. Field experiments are made but some have served primarily consensus building functions. In no case did the final decision to introduce videotex actually depend on the results of a public trial. Conflict about and opposition to the planned introduction of videotex came to the fore as the field tests demonstrated possible modes of utilization. There is least evidence of opposition in the British case. In the two other countries, opposition died down after some appeasement measures, and even before that it never reached a particularly high intensity. Powerful organized interests such as the labor unions did not appear to be negatively affected to any significant degree, and while the print media did fear negative consequences, they had also good reasons not to oppose this technological development outright. It is interesting to note that videotex has not become a partisan political issue in any of the countries: though in all three cases important government changes took

Table 1: Comparative Chronology of Videotex Development

	France: Teletel	UK: Prestel	FRG: BTX
1971	Technological developments at the PTT's major research center CNET.	Technological developments leading to Prestel at the UK Post Office research laboratories;	
-74			
1975		Post Office presents "the world's first videotex system" to the public.	
1976			Government advisory commission KtK recommends attention to videotex; Bundespost decides to develop videotex.
1977	DGT, the French PTT's telecommunication division, is alerted to telematics progress in other countries; prototype of Minitel is developed.	Prestel trials start; about 100 information providers.	Purchase of licence for the British Prestel by the DBP; BTX is shown at at International Fair in Berlin
1978	Nora-Minc report published; Government decides experimental introduction of electronic telephone book and of experimental videotex service on adapted TVs.	Formation of Association of Viewdata Information Providers.	Establishment of a Working Group for BTX at the DBP with producers, IPs and users; non-public testing; controversy about legal status of BTX.
1979		Prestel is launched as a commercial service before the end of its market trial; TV manufacturers welcome the chance of a pre-developed service from the Post Office; Prestel-adapted TVs envisaged as main terminals.	
1980	First test of "annuaire électronique" at Saint Malo; DGT announces plan to introduce Teletel; core service will be annuaire électronique; private service providers may join.	Prestel marketing strategy shifts from "domestic" information services to professional services.	Start of field tests in Berlin and Düsseldorf-Neuß; 2000 private, 1000 professional users participate

France: Teletel	UK: Prestel	FRG: BTX
1981 Opposition to telematics program especially from press; commission is set up to evaluate the telematics experience; DGT orders 300,000 terminals; Second test opens in the Rennes area; Field test is started Vélizy; New minister of PTT announces revised telematics program.	Prestel registers its 10 000th user nationwide; introduction of Prestel's messaging service; access to Prestel from seven countries becomes possible; travel agents become important user group.	Cabinet decides that BTX will be introduced; CEPT standard is adopted; IBM Germany gets commission for BTX systems equipment.
1982 Service starts in Nantes, Grenoble, and Strasbourg, nationwide professional Teletel started; another 300,000 terminals ordered.	British Telecom (BT) separated from the P.O.; creation of BT Enterprises of which Prestel becomes a part; Prestel opens gateway service permitting access to non-Prestel computers; launch of Telecom Gold (a competing service) within another division of BT.	PTT minister announces that DBP wants to develop BTX into a mass service.
1983 Regular service starts in "Picardie" and "Ile de France".	Reorientation of Prestel to the mass market with new transaction services Homelink (telebanking), and Micronet (service for home computer owners); BT reorganized; Prestel and Telecom Gold (a competing service) belong to the same divison.	DBP fixes BTX charges; Convention concerning BTX signed by Federal Government and Länder, BTX officially inaugurated; six months late due to delays in system installation. New IBM system put into operation.
1984 Legal regulation of service provision; New tariff system "Kiosque".	Launch of the Prestel Education service; British Rail provides ticket booking service on Prestel.	
1985 Spread of many message services; breakdown of the Transpac Packet Switching System due to overload.	Telecom Gold has almost as many users as Prestel.	Siemens' bitel (BTX-telephone) gets licensed; DBP develops new marketing strategy, oriented toward small businesses.
1986 Further simplification of access for service providers.		DBP launches program to lease BTX compact terminals (multitels); Electronic telephone book is introduced.

place (Mrs. Thatcher came to office in 1979, Mitterand in 1981 and Kohl in 1982), this did not result in any radical changes with respect to videotex development. In the UK, however, the conservative liberalization policies had far-reaching long-term effects by leading to early and intense competition between *Prestel* and a growing number of similar services.

Phase 3: In this last phase, the planned nationwide system went step by step into routine operation. At this point, commercial service providers, a range of user groups, the associations which both formed, and - in the British case - competing technical systems entered the scene and shaped the growing system. In two of the three countries, expectations of mass utilization were disappointed, which led to the adaptive modification of some systems features and changes in marketing strategy. The process of videotex development is still incomplete; in none of the countries has a "momentum phase" been reached, and it is not even clear that this will generally happen. In fact, the monolithic videotex systems which we can today observe in France and Germany may disappear in an array of overlapping and competing services, as seems already to be happening in Britain.

After this brief introductory description we shall analyze the features of the developing videotex systems and their utilization in more detail. We shall then try to explain the observed differences, and will conclude with some reflections on features of the development process.

3 Cross-national differences in systems design and user patterns

The initial idea of videotex was to have low-cost and user-friendly access from user terminals to computer centers via data transmission facilities in telecommunication networks. There are several technical ways in which this idea can be realized. Thus, a videotex system may consist of a distributed network of independent computers, a hierarchy of computers with external data bases, or a mixture of both. A variety of options also exist with respect to the transmission technology and terminal configuration. Table 2 summarizes the major design alternatives, but it also shows that all videotex systems are composed of certain basic elements: a host computer (or set of host computers), a network to access the host(s), a terminal on which the text is displayed, a display standard used to define the character set and the graphical signs communicated through the system, and a retrieval system facilitating the access to information also for non-specialists.

Table 2: Structural Elements of Videotex Systems and their Combination

Terminal Configur.	Modem	Telecom Networks	Databases	Standards
TV+Decoder PC+Hardware Dec. PC+Software Emul. Profess. Terminal Multifunctional Videotex Telephone	Modem	Telephone network and/or Special data network and/or Packet Switching network	Central and/or regional and/or local Computer centers and Databases	Prestel Antiope Telidon CEPT Captain (ASCII) .. .. etc.
Integrated Compact Terminal (French Minitel)				

An important feature of videotex is that it is not designed for one specific form of utilization only (as is the telegraph, for example). It is rather a communication infrastructure which permits a number of different forms of usage, depending on the terminals, display standards and network architecture which have been chosen. Possible applications range from message systems and information data bases to transaction services such as home banking, home shopping, etc. Applications also differ with respect to the target groups to which they are oriented and by which they are mainly used - private households, professionals or business firms.

In view of this plurality of design options it may not be surprising that the videotex systems developed in Great Britain, France, and Germany differ markedly from each other. This fact *per se* reflects an insight which today need no longer be argued at great length, i.e. that the design specifics of technical systems are shaped by social - economic, legal, political, cultural - factors. But where choice exists, the selection of specific alternatives needs to be explained, and this we set out to do in the following sections. In this section we shall first describe the three different videotex systems and the - similarly different - utilization patterns which develop in response to the opportunity structures thus created.

The technical system and its social organization

The technical and social structure of videotex systems may be described by specifying four core elements:

- the videotex actors, i.e. service providers, system operators, communication network providers and producers of hardware and software
- the technical architecture of the system: database arrangements and communication networks; the display standards; the typical user terminals; the methods and procedures of accessing the services and information (retrieval systems) etc.
- the organizational structure: rules and regulations which allocate tasks and responsibilities such as: system operation, hardware and software provision; administrative and control activities including passwords, user access, billing, central indexing, messaging and any form of user monitoring
- the regulatory norms: although sometimes difficult to distinguish from the previous category, the regulatory norms refer to rules related to externalities such as privacy and data protection, consumer protection and other "social control measures" in the innovation process.

Actors

There is no basic difference between the three countries in this respect: In all countries the system operators are the PTTs, which also provide the communication networks. In all three countries there are private information or service providers, and in all three countries private enterprises produce the hardware and software components for the system. So far the actor systems look similar. However, the industrial producers in France come mainly from the telecommunications domain, whereas in Germany and Britain they initially came from the consumer electronics sector.

The technical structure

As outlined above, on a rather simplified level the technical systems can be described by specifying the terminal configuration, communication networks, database arrangements, and display standard. For the communication network there is only a limited room for choice. The existing videotex systems typically involve at least two network linkages: one between the user and the videotex service centers, the other among the service centers. In all three countries the telephone network is used for the first linkage. For the second linkage France chose a special solution because it uses the general public packet switching network for this purpose, while the two other countries use a special data network.

Regarding the data base arrangements, the German and British systems are rather close: both have very centralized network architectures. The *Prestel* system maintains a master database at a centralized update center and replicated databases in a handful of information retrieval centers. The data base is therefore replicated, rather than distributed, across a number of machines. This creates a demand for large storage capacities[9]. In the original version of *Prestel* there was no gateway for connecting remote databases to the system. Consequently it was not possible to implement true interactive applications such as, for instance, telebanking. This facility was created when the *Prestel* system was modified in 1982 and the German "remote database network" concept was applied.

The core of the German system is a highly complex hierarchical system of databases and computer networks which was designed and implemented by IBM. A recent study of an international consulting agency (Butler Cox) called the German system the most complex and sophisticated system in the world[10]. The central strategy of IBM was to create one single big database and network management center together with a stratum of regional databases in which only the most frequently used information pages are stored. This favors updating from one single center and reduces the overall storage capacity required[11]. Another special feature of the German network architecture is the possibility to connect remote databases to the system via the packet switching network. For this connection, however, a very complicated communications protocol is required in addition to the well-known X.25 standard, which makes access rather difficult and costly for information providers.

In contrast to the videotex systems outlined above, the French system has no central database and no specific network infrastructure beyond the already existing public telecommunication networks: The French *Teletel* consists of a large number of autonomous, privately-owned computers (servers) which are interconnected by the public packet switching network and can be accessed by the telephone network. The most striking difference between *Teletel* and the other systems, therefore, is its complete decentralization. Unlike the British and the German systems, *Teletel* has no central database provided by the systems operator. Subscribers are connected via the telephone network to a switching computer which works as an access point of the public packet switching network *Transpac*. This data network then links the subscriber to a remote database, which is chosen by the special code of a service. This systems architecture has far-ranging implications for the flexibility of adapting to the users' changing needs and for the financial access barrier for service providers.

A significant difference between *Bildschirmtext*, *Prestel* and *Teletel* exists also within the (typical) terminal configuration and with respect to the display standard. In Germany and Britain it was initially thought that the television set enhanced with a special decoder should be used as a display device. The television industry was expected to exploit this chance for new markets and to develop cheap decoders which would lead to low financial access barriers into the system. Since in both countries the "rush" of private households has not yet taken place, today's typical terminals are not television sets. The most widely used terminals today seem to be professional terminals, which are manufactured exclusively for using *Prestel* or *Bildschirmtext*, or personal computers - now that relatively inexpensive microcomputers are generally available. The French terminal configuration differs completely from that of the Germans and the British. The typical *Teletel* terminal is the *Minitel* - a very simple compact terminal equipped with a small monochrome monitor, a modem, a decoder and an alphanumeric keyboard.

Further differences between the three systems exist in the display standards. Despite their seemingly marginal importance, these technical aspects have an important impact on the complexity of the hardware and software requirements for the decoder. The basic differences are represented in Table 3. As this table shows, the *Prestel* standard is the lowest - because compatibility with the British broadcast-videotext played an important role in its design when it was developed in the

1970s. However, the impact of the screen standard on speed and storage seemed to be also an important criterion in the British Prestel design.

The most complex display standard was developed at the CEPT level. This standard is used within the German system[12]. Although this is a European norm, standardized at the *Conférence européenne des postes et des télécommunications*, it was strongly influenced by the German Bundespost. The functional requirements for the hard- or software which are implied in this standard are so intricate and complex that even a multinational corporation such as Philips was unable for several years to realize these specifications in an Integrated Circuit (IC) without major difficulties.

Table 3: Technical Differences in Display and Transmission

Specifications	Prestel	Teletel	CEPT (Btx)
Resolution	6x10	8x10	12x10
Transmission Speed	1200/75 bit/s	1200/75 bit/s	1200/75 bit/s
Page Format	24x40	25x40	24x40
Characters	95	127	335
Graphic Symbols	64	64	151
DRCS*	-	-	94
Colors	8	8	4096
Coding	7-bit	7-bit	8-bit

*A DRCS is a character whose shape is freely definable.

The French system steers a reasonable middle course between "technical performance" or sophistication on the one hand and "financial burden" on the other. The *Minitel* has another great advantage: Its screen resolution is based on 8x10 matrices and the 7-bit coding which are well known in the home and personal computer domain. The French display standard is thus more closely related to common computer technology. As a consequence, it is technically less difficult to implement the ASCII and 24x80 character mode (the normal professional database standard outside the videotex) in the French *Minitel* than in the other systems. The adaptation of *Bildschirmtext* to normal database standards is technically far more difficult to realize. Although *Prestel* with its 6x10 resolution is also incompatible with the home and personal computer standard, the low complexity of its standard nevertheless renders similar solutions easier. On the British computer software market there are now many terminal emulation and communication programs available in which the *Prestel* facility is implemented together with other communication protocols.

Organizational structure

One basic organizational feature of videotex is the distribution of the responsibilities for the system's operation among different actors, e.g. are the system's operators private or public - or both? In this respect there are important differences between the three systems.

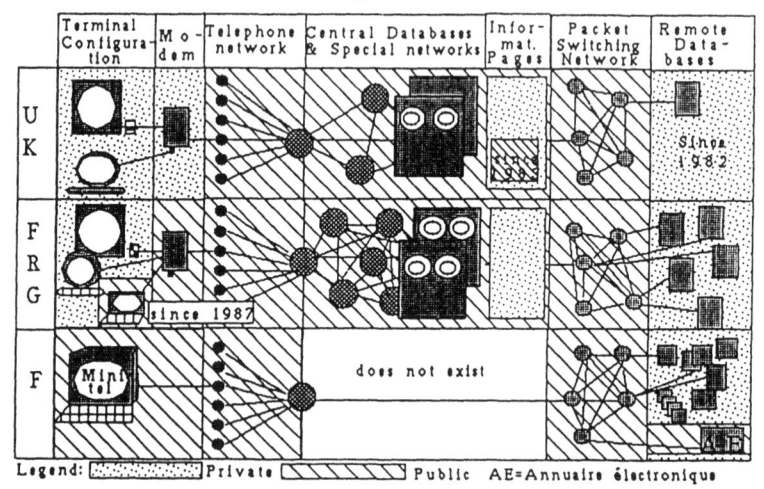

Figure 1: Technical and Organizational Structures

Because of the telecommunications monopoly which existed during system design in all three countries[13], the postal organizations generally provide the networks and switching facilities and, thus, the basic infrastructure of videotex. Nevertheless, a major difference between the French and the other two systems is that, in addition to controlling the telecommunication network, the French state also controls the terminal sector and, with the electronic telephone book service, one important service. Therefore, within the French configuration at least for one application the whole videotex *filière* is controlled by one single actor: the French PTT. This solution could be called a "state solution". However, the service sector beyond the electronic telephone book is a complete market solution in France - a market which quickly passed its critical mass and is currently growing very fast.

The German and the British systems have adopted a different distribution of tasks between private and public actors: in both countries the PTTs are only responsible for the storage and transmission of information, the information providers are responsible for information contents, and the terminal market is under the control of private firms.

A further element of the social organization of videotex systems are the regulations concerning the financial contributions of information providers and users. In Table 4 the user tariffs and user tariff structures of the three systems are outlined. As can be seen, the British and German systems have a similar tariffication policy, whereas the French system is unique in the sense that there are no standing charges at all and all charges are time based. The most important French tariff subsystem is the *kiosque*: its time charges cover at the same time transport *and* service costs.

Regulatory norms

The major regulatory aspects of videotex concern equity of access, consumer protection, and data protection. Comparatively speaking, the German *Bildschirmtext* is the most heavily regulated videotex system. The German terminal market is subject to a dense network of regulatory constraints, especially licensing procedures. To give an example: In the interest of consumer protection, the German system requires the *Bildschirmtext* user who calls an information page for which there is a charge to confirm his intention of looking at the chosen page by typing the corresponding numerical code of "yes". In sharp contrast, data and consumer protection is almost non-existent in France, and access for information providers is practically unconstrained[14]. This may create a number of problems in the future, but it undoubtedly has

Table 4: Tariffs and Tariff Structures in Comparison

	Prestel	Teletel	BTX
Standing Charge	residential users £6.50 per quarter (~6.50 DM/month) business users £18 per quarter (~18 DM/month)	no standing charges	8 DM/month
Time charge	Telephone charges 51p/h during the cheap rate period (~1.50 DM/h) Special Prestel Charge during telephone peak hours £3.60/h (~10.80 DM/h)	all teletel charges are time based - electronic telephone book: the first 3 minutes are free; then for every 2 minutes 0.75 FF; for the other services there are 3 different tariffs: - "3613": 0.75 FF every 6 minutes - "3614": 0.75 FF every 2 minutes the service provider then sets an additional time charge. - "3615" (Klosque-System): 0.75FF every 45 seconds. This covers transport and service costs (~24 DM/h)*.	telephone charges: every 8 minutes 0.23 DM (during cheap rate periods every 12 minutes 0.23 DM). (~ 1.70 DM p/h)
Frame Charges	frame charges are set by the information providers; most pages are free.	no frame charges: all charges are time based	prices for frames are set by the information providers; however, there is a maximum of 9.99 DM/per frame.

* since the end of 1987 there are several additional, more differentiated tariff categories within the Klosque

the advantage of facilitating and smoothing the videotex usage both for service providers and for users.

Evolution dynamics, diffusion and user patterns

The various features of the socio-technical system "videotex" constitute an opportunity structure with incentive and disincentive effects for different groups of potential users, whose reactions in turn constitute an opportunity structure for service providers and hard- and software producers. The interdependence between provision and utilization operates as a positive feedback loop which can generate dynamic growth, but which may also lead to a downward spiral of decreasing utilization and decreasing service quality, unless countermeasures are adopted.

The difference in the growth dynamics of the three videotex systems can be described with the aid of various time series. The development of the number of user terminals may indicate how the systems grow and how the videotex services find acceptance among users. The figures on the growth of information or service providers may give an impression of the development of the new "telematic market" (databases, communication services, transaction services etc.). Table 5 and especially Figure 2 show a striking difference between the French, German and British videotex systems with respect to the number of terminals connected to each system. Whereas the British and German figures with a more or less stable growth rate look rather similar, the diffusion of French videotex terminals has grown exponentially during the last three years. Compared to the British and German "failures" or "flops"[15], the French, indeed, are "riding a videotex craze"[16], and French PTT managers are peddling their "success story" internationally[17]. The growth and structure of service providers or information providers in the three countries cannot be compared directly because the systems are differently structured. Their evolution over time, however, may give a rough picture of the different growth dynamics in the three countries, in particular in the service market. Whereas the number of information providers in Germany and Britain is more or less stagnating or even decreasing, the number of services in the French system is growing very fast.

Table 5: Diffusion Patterns in Videotex: France, Britain and the FRG

	1982	1983	1984	1985	1986	1987
France: Teletel						
Subscribers (end of year)	-	120,000	530,000	1,300,000	2,200,500	2,791,000*
Services (end of year)	-	-	840	2,070	4,150	5,662*
Hours of connection (mill.)**	-	-	-	1.0	2.5	4.7*
Britain: Prestel						
Subscribers	19,850	38,000	48,000	63,000	70,000	76,000
Information providers	1,003	1,356	1,365	-	-	1,252*
Number of frames	41,050	277,100	330,000	320,000	300,000	310,000
Frame calls per week (mill.)	-	-	3.4	7.6	9.1	9.1
FR Germany: Btx						
Subscribers	-	10,155	21,319	38,894	58,365	83,633
Information providers***	-	2,740	3,099	4,043	3,528	3,416
Remote databases	-	0	37	151	218	248
Number of frames	-	378,000	521,783	762,673	589,330	610,704
Number of calls (mill.)**	-	0.1	0.3	0.5	1.1	1.9

* Data refer to June 1987
** per month
*** Sub-information providers are included

Against this background it is clear that the French videotex system *Teletel* has had the most spectacular development. By the end of 1987 almost 3 million *Minitels* had been distributed by the French DGT. The number of accessible services tripled in the course of two years, from about 2000 services to about 6000. This growth continues - every day several new services are created. A large part of them are offered within the tariff system called *kiosque* which provides the easiest and most flexible access through its time-based billing system. The services within the *kiosque* are in the greatest demand. From the 3.5 million hours of connection in the year 1986 (not counting the electronic telephone book), about 2.5. million were within this category.[18] The structure of the service providers is quite diversified, but services oriented toward the general public are in the majority.

The overwhelming share of use of the French system is non-professional in nature; professional demand, nevertheless, is also quite strong. There are data which show that of the 1,300,000 Minitel terminals distributed by December 1985, nearly 40% were installed in firms.[19] Other figures similarly suggest that Teletel has established a strong foothold in the business community.[20] Contrasting sharply with the French *Teletel*, the terminal diffusion and the applications of British and German videotex are largely limited to professional users - this is evident also with respect to the service structure (cf. Table 6).

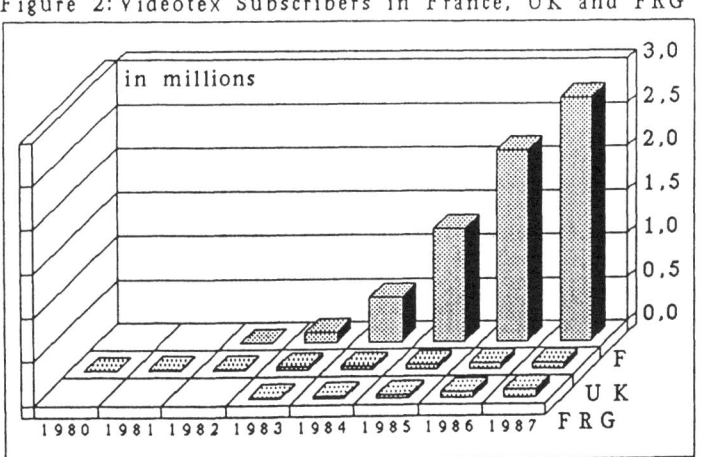

Figure 2: Videotex Subscribers in France, UK and FRG

Although for Germany there are no reliable figures about the distribution of terminals among professional and residential users, it has been estimated that only about 20% are privately used.[21] This is more or less also true for Britain's *Prestel*. Although British Telecom as *Prestel's* operator publishes exact figures on the distribution of *Prestel* subscribers between the private and professional sector, (according to these data, in summer 1987 about 39% where used at home[22]), these figures seem to be unreliable. Insiders say that a large share of the terminals which have been classified as private are in fact used for professional purposes[23].

The slow growth of subscribers and the marked reluctance of private households to use and pay for the services is accompanied by a relative quiescence in the arena of information and service providers. In Britain and in Germany there are only very few services which are profitable. The overwhelming majority of service providers are making losses and are only staying in the system for strategic purposes. It is still expected

Table 6: The Service Structure (in percent)*

Branches	UK ('82)	FRG ('85)	France ('86)
Press, Media, Communication	14	11	35
Other "general public" services	-	-	29
Tourism	32	4	6
Trade, Electronics	7	18	-
Finance	7	24	12
General industry	-	5	9
Consulting firms	13	15	-
Public institutions	11	14	6
Education	6	-	3
Miscellaneous	10	9	-

* These figures are only roughly comparable, since the statistics - esp. F compared with FRG and GB - work with different categories; updated figures for the UK are not available.

Sources: Minitel Guide des Services; ISI 1987; Btx Praxis (1985)12; Butler Cox (1981/1982)

that - in the long run - a market for telematic services will evolve.[24] Especially in Germany this has led many business users to use videotex as in-house networks and as cheap data communication facility between firms.

In conclusion, only in France has the evolution, diffusion and usage of videotex fulfilled the optimistic initial expectations. In contrast to Germany and Britain, France has succeeded in penetrating also the private household area with this technology. In doing so, it has created a very dynamic market for telematic services. That this "success" was possible is undoubtedly a function of deliberate choices in the technical and organizational design of the system, but it must also be related to contextual and situational factors which explain why some choices were possible and others were not.

4 Explaining the cross-national differences: actor strategies, technical opportunity structures and institutional arrangements

In trying to account for the cross-national differences in videotex development outlined in the previous sections, our framework of analysis starts with a structured system of actors who, under given environmental conditions such as institutional and resource constraints, shape a technical system. This system in turn creates an opportunity structure for a set of applications. Realized applications then shape the usage patterns.

A central presupposition of this approach is that differently structured actor systems together with different actor strategies lead to different technical systems configurations. Actor systems and strategies must be explained within the context of particular economic, cultural and legal arrangements which enable some actors to act or "think" in a certain way. The structural constraints under which they operate should not be seen as deterministic effects of external variables, i.e. economic, institutional and cultural factors, nor should these factors be conceived as being static over time. The constraints are transmitted and reproduced in each action situation, and they can vary as the result of voluntaristic utilization and extension of the room for maneuver of the actors involved. Our basic explanatory idea is therefore that country differences in contextual conditions and in the actor systems

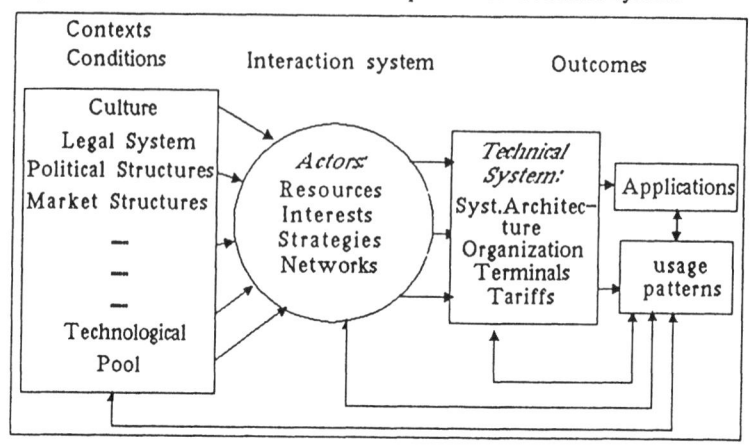

Figure 3: Determinants of the Development of Technical Systems

Contexts Conditions	Interaction system	Outcomes
Culture Legal System Political Structures Market Structures — — — Technological Pool	*Actors:* Resources Interests Strategies Networks	*Technical System:* Syst.Architecture Organization Terminals Tariffs → Applications ↔ usage patterns

involved in videotex development can explain the different national technical system configurations and the application (usage) patterns of the videotex technology.

Core factors explaining the observed differences between the three videotex systems are found (1) in the general *introduction strategies* (or policies) adopted by the three PTTs, and (2) in a number of *key decisions in the technical and organizational design* of the videotex systems. Whereas the introduction policy refers to procedural decisions, cooperation strategies of the dominant actors and the distribution of competences between private and public actors, the design decisions are more related to the specific attributes of the technical and organizational system configuration itself.

Introduction strategies

The introduction strategy for *Bildschirmtext* and for *Prestel* follows the logic of an infrastructure policy orientation, mediated by the self-interest of the PTTs. In both countries, the main motivation of the PTTs was to create a new growth field within the telecommunications domain because telephone diffusion had reached its saturation level. In addition, they hoped to stimulate the use of the telephone network, especially in off-peak times. In Germany the Bundespost was also looking for investment opportunities for its profits, which the Federal Government otherwise might have absorbed into its general budget.

A major determinant in the British and the German introduction strategy was the terminal technology available. As terminal display devices were still expensive during the 1970s and the PTTs were looking for a new mass market, the revolutionary idea was to use the home TV, with which 90% of the private households were already equipped, with a special adapter. This technological strategy implied that the PTTs would have to rely on the consumer electronic industry instead of entering the terminal market themselves. Given the difficulties which this industry had with Japanese imports, it was supposed that the firms would be interested in the new market.

A further element of the introduction policy in Britain and in Germany was the "common carrier" idea, which meant that the PTTs would provide only the technical infrastructure for storage and transmission. Consequently, the development of the videotex service domain was left to be driven by market dynamics. This meant that the growth of such a market depended very much on the number of service subscribers. Both these countries thus faced a critical threshold problem which led to later strategic modifications. British Telecom as *Prestel* operator gave up the "common carrier" concept in 1983/84 and is now offering services, too, while in Germany emphasis was shifted from domestic to professional users.

The French strategy of videotex introduction differed from that of Germany and Britain in two important respects. First, the introduction policy was not based on a mere telecommunications infrastructure policy primarily oriented toward assumed societal needs, but on a voluntaristic sector-specific industrial policy which aimed to create new markets in order to develop industry. Secondly, videotex was not primarily introduced to create a new telecommunication service, but was "sold" as an internal postal rationalization project, in which the rather costly operator-assisted telephone information and the printed telephone book was supposed to be replaced by the *annuaire électronique*. This justified, in fact even required the distribution of *Teletel* terminals free of charge. In following this strategy, France had learned especially from the British mistakes. It is striking how clearly these lessons were spelled out by Roy D. Bright, a *Teletel* manager who was a former *Prestel* manager, in 1982. Under the subtitle "the videotex learning curve", Bright spelled out the following points, that could be learned from the British experience: "(i) The reluctance of the mass market user to bear a major proportion of the cost of the service. (ii) The lack of commitment from TV manufacturers while the market is still in its infancy resulting in high terminal costs. (iii) The danger in creating a centralized system which cannot readily adapt to the various needs of different 'service providers'."[25]

Based on these insights, the French PTT decided to assure a fast diffusion of videotex terminals, and the easiest way to do this was their provision by the state free of charge. In connection with this, the French PTT offered with the very smoothly functioning electronic directory service at least one strong user incentive. Private operators were then free to create additional, independent services. Following this strategy, *Minitels* were distributed from 1982 on in all French departments. The assumption was that these investments would be written off within 7 years merely by the increase of traffic within the telecommunication networks. This strategy succeeded at least in one respect: with a huge investive advance, the French obviously passed the critical threshold and, in the meantime, created a very dynamic market for communications services and information.

The French videotex introduction strategy resembled very much the traditional French mercantilistic orientation where the control of industry is used for the achievement of political goals.[26] In fact, within the tradition of the French "grands projets" the *annuaire électronique* was used as an instrument of a general industrial policy, which was geared to challenge the American hegemony in the field of information technology[27]. In this connection the French policy makers even have developed the strategic concept of the *filière électronique*, which essentially tries to identify strategic sectors within the system of technological and sectoral interdependencies. The key idea there is to develop weak sectors with the aid of closely connected strong sectors.[28]

Choices in systems design

The introduction strategy of the French planners goes a long way toward explaining the current success of *Teletel*. Nevertheless, there are also some technical and organizational decisions which, although

sometimes tightly coupled with the overall French introduction strategy, can be treated as separate choices. In the technical and organizational systems design the French government made several strategic decisions which had important consequences for videotex utilization:
- The first was the technological choice to produce a very simple terminal which would be easy to handle, and cheap.
- The second choice was to establish a highly decentralized and flexible network structure, which could easily be adapted to changing user needs.
- The third choice was to create a very unbureaucratic billing system, which, on the one hand, unburdened the service providers from costly administrative work and, on the other hand, enabled a free user access without any formal administrative entrance barriers.

The French choice of a simple terminal may have been dictated primarily by financial considerations, as terminals were to be paid for by the state itself. But it is not only the low financial access barrier to the user which promoted videotex growth in France. With its simple and commonly used communication procedures (X.25) and its simple display standard, the French system also made it easier (and less expensive) to develop truly interactive services.

The decentralized French network concept reinforces these effects. The absence of a central database in France stimulated the development of transaction services as service providers and users interacted directly in any case. In Britain it took years before it became possible to connect external computers to *Prestel*. In Germany, the possibility to do so existed from the beginning, but it is still an expensive and complicated alternative to the use of the central databases with their limited interactivity.

The flexibility of the French decentralized system also facilitates service innovations. An example for this is the invention and development of interpersonal communication services. What today is called a *messagerie* was invented as a result of a system failure in a videotex field trial in Strasbourg. Once the technical possibility of anonymous communication was discovered, the idea was taken up by some service providers, who were able to implement this development in their own host computer. The readiness of the French PTT to permit this new form of communication then led to applications which the system builder had not thought of.

The French decentralized technostructure also much better fits the requirements of updating. A number of field trials and many market research studies have shown that the strength of electronic communication lies in providing *up-to-date information* - and being continuously up-to-date depends on a direct and easy access of service providers to the host. It is interesting that only the French system really supports applications where up-dating is no important financial and technical problem, whereas in *Bildschirmtext* and *Prestel* such applications are technically much more cumbersome. In consequence, the majority of information providers use the support of consulting agencies specialized in such services. This makes updating in *Bildschirmtext* relatively expensive in terms of time and money. As a result, only a very small group of information providers update their pages continuously, which lowers their attractiveness in many cases[29].

Given the intentions to create a mass service, the German and British choice of a centralized solution seems paradoxical - but the choice is understandable within the context in which it was made. It was a time in which microcomputers did not exist and electronic data processing still had high financial entrance barriers. Especially interested groups representing small business therefore supported the Bundespost in the establishment of a centralized public database. It was thought that if a public storage facility did not exist, only big firms could establish remote databases and small business would be excluded from the information market. But as the central database turns out to be a major hindrance for the provision of truly interactive services and to be too costly with respect to updating procedures, it is now in effect the centralized system which excludes the small information providers from key applications of videotex.

Many observers consider the French *kiosque*-billing system the most important decision in French videotex system design. It is generally believed that the dramatic increase in videotex traffic was due to the introduction of the *kiosque* facility for general public-oriented services in 1984[30]. The completely decentralized French network structure originally implied that each information provider would have to create his own billing system. Except for the free services the user would then have been constrained to subscribe to each individual service separately. But with the *kiosque* system a general time-based billing system was introduced which saved information providers these administrative efforts. In the *kiosque* system each service is accessible without password and costs the user about 1 franc per minute. The French PTT collects the charges with the normal telephone bill, keeps 3/8 of the total charge for transport, and sends 5/8 to the service provider. Interestingly, many users prefer to access services via *kiosque*, even if the same services are cheaper to access by subscription. The success of this billing system may largely be explained by the fact that it responds to new trends in consumer behavior: the aversion to constraints which are created by formalities of subscription and the interest in freely passing from one application to another.[31]

The design of the German and British billing systems are very much determined by the "paging concept" coupled with the common carrier idea. The basic idea there is that the costs of operating the system should be covered by standing charges for users and storage charges for information providers. The users then have to pay the information providers separately - page for page - unless a service is offered free. In this system, the user is constrained to consider the "price-for-service relation" continuously as he uses different videotex services.

A factor which is related to the organization of the system and which hindered the dynamic growth of innovative applications in the German system has been the regulatory overprotection of *Bildschirmtext*. Although this was not a deliberate choice of a single actor but the joint result of the actions of a set of actors, it has to be considered a strategic decision, too. As the Bundespost initially relied on the TV as the display terminal, *Bildschirmtext* was conceived by the public as a new electronic mass medium. This triggered the intervention of media policy actors who wanted to incorporate this service into their domain of regulatory responsibility. *Bildschirmtext* thus became the most intensely regulated videotex service. The relatively low regulation of the French system is an important factor which should not be underestimated when explaining the dynamic growth of the French service market and the rapid diffusion in the private sector.

Environmental conditions and institutional contexts

We have shown that success and failure of the different videotex systems have to be explained to a large extent by strategic decisions of the core actors on the one hand, and by the resulting techno-logic of each system on the other hand. These explanations, however, are incomplete. The strategic actors' decisions cannot be taken for granted. Strategies themselves are embedded in a structural and institutional environment. It is necessary therefore to understand under which circumstances and in which context these decisions were feasible, and which other options were systematically ruled out. It is therefore necessary to look at the particular institutional and cultural background of each decision.

The most important question would be why the Germans and British did not apply the same introduction strategy as the French when they launched their videotex systems. Here the built-in restriction in the concept to use the existing TV terminal is very important. The result was that the British and the Germans from the very beginning relied on the consumer electronics industry, an industry in which public purchasing played only a marginal role. The French strategy, in contrast, was dominantly based on the telecommunications industry. Since the telecommunications monopoly in this domain implied a long tradition in public purchasing with a well established network of "court suppliers", the *Minitel* strategy was just the continuation of traditional business.

Further pressure against a telecommunications based introduction strategy existed in Germany because since the late 1970s there was a growing resistance on industry's part (especially by the computer industry) with regard to the PTT monopoly in the terminal market. It would have been unthinkable that the PTT would have been allowed to provide the *Bildschirmtext* terminals following the old telephone strategy where markets had been reserved for some court suppliers.

The choice between telecommunications industry and electronics industry had important consequences. Both sectors have completely different market structures and are organized in different ways. In all three countries - at least until 1984 - the telecommunications industry was highly concentrated and protected from external pressure; procurement relationships were stable and restricted to a small set of traditional court suppliers. The telecommunications market showed a "clientelistic structure". The consumer electronics industry in contrast is very dynamic and open to foreign competition. Even though this sector is also relatively concentrated and penetrated by conglomerate corporate structures, the international openness creates a highly competitive market. In contrast to Britain and Germany, whose PTTs had to negotiate with more than a dozen TV manufacturers, the French administration was able to procure its *Minitels* from a small group of traditional telecommunications firms and was therefore able to exploit large "economies of scale". The Germans and the British also had to deal with the fact that the TV manufacturers were less interested in producing "external decoders" to adapt existing TV sets to videotex than in fostering the TV replacement cycle with videotex, understandably so in a saturated color TV market[32]. British and German TV manufacturers consequently were doing little research and development[33]. Also, too many competitors in a sunrise market prohibited the expected price decrease for terminal equipment. This is still considered to be the major hindrance to the videotex boom. This "crowding effect" is further increased by the merging of home telematics with conventional office information and communication systems. One consequence is that computer firms are becoming more and more interested in the terminal market. The intrusion of the dynamism inherent in the computer sectors increased the openness and competition in the German and British terminal markets even more. In such a context, it was hardly possible to apply the traditional telecommunications terminal provision strategy. But situations and contexts can change. The fact that the German PTT, in a strategic reorientation with its "Multitel Program", is now trying to follow the French strategy, and that industry is not protesting is tacit recognition that it has failed to produce a functioning market. Interestingly, even the deregulated British Telecom now intends to "go Minitel"[34].

The French videotex strategy is closely linked to its deliberate reliance on the telecommunications industry. At the same time, the French strategy is also very much related to the French historical tradition and institutional framework. The governmental planning system, headed by the *Commissariat Général du Plan*, has long been a key instrument for channelling large capital investment into selected economic sectors.[35] Especially in telecommunications, this is supported by close links between the Ministry of PTT and the telecommunications and electronics industries. In addition, there are also close organizational links between the French PTT and the *Ministère de l'industrie*. The PTT has an office of industrial policy which can be used for coordination with the more general industrial policy of the government[36].

Important for the explanation of the French strategy is also the special situation of the French electronics industry in the 1970s. Beginning in the late 1960s, France tried to bridge the "technological gap" to US computer industry. But the great *Plan Calcul* ended unsuccessfully in 1974. The French telecommunications sector, in contrast, was remarkably successful during the 1970s. Thanks to public procurement policies guaranteeing long-term orders and adequate research funds, the French telecommunications industry has been able to develop very advanced technologies and modernize the French telephone network in a very short period. In the course of one decade, the French PTT quadrupled the number of telephones from 5 million in 1970 to 20 million in 1980 and succeeded in developing the world's first fully-digital exchange and packet switching network. With the *plan télématique* the French government tried to repeat this success in the new domain of "telematics". The aim was to use the telecommunications industry as a "lever" to create a telematic sunrise industry.

In Britain and Germany, where state intervention has always been a rather contentious issue, PTT procurement policies certainly shape industrial decisions as well. But in both countries the PTTs have never been used as policy instruments in order to pursue general industrial policy objectives. In Germany there is no political infrastructure for instrumentalizing the Bundespost for industrial policy. Institutionally and financially, the Bundespost is an almost completely autonomous organization, and its investment strategies and goals are derived from its own preferences and objectives. Key decisions made when *Bildschirmtext* was introduced are good illustrations of this general orientation. The first decision was to buy the *Prestel* system for field trials. At this time, the German military firm Dornier, with the aid of the German Ministry of Research and Technology, was developing its own videotex prototype. Surprisingly, the Bundespost did not even take this national videotex project into account and bought the foreign *Prestel* system. The other decision was to charge IBM with developing and implementing the public service centers. IBM won a public tender, although SEL, a traditional German "court supplier", had also made an offer. Within the French industrial policy orientation, such a decision would hardly have been possible.

Compared with France, there is certainly nothing like the French "industrial planning system" in Germany and Britain. Industrial development in both countries is much more an effect of market forces and the strategic decisions of the big firms than of state policies. Although there are some sporadic examples of high-tech industrial policy (e.g. the British teletext initiative[37]) and there is much governmental rhetoric in this domain[38], consistent sector-oriented state intervention in

Germany and Britain exists only in "sunset" sectors. But even in the German steel sector this interventionism is currently hotly debated. Such restrictions for consistent and strategic state interventionism follow from the hegemony of a "liberal market ideology" in both countries. Closely related to this is the "balanced budget" philosophy, which meant for *Bildschirmtext* and *Prestel* that both systems were designed to be financially self-supporting.

Another strategic decision which has to be explained is why the Germans have chosen such a complex display standard. Although in the face of a primary orientation toward the home market the preference for high graphic capabilities seems understandable, it is still hard to explain why the Bundespost pushed for such an overcomplex standard. Foreign observers do so by reference to German techno-perfectionism. There seems to exist a techno-culture in Germany which emphasizes "functional sophistication", technical finesse and "overengineering" regardless of the technical and financial burden which this involves. The French and the British seem to be much more pragmatic in this respect.

The overregulation in the German case finally is linked to the federal institutional structure, which distributes regulative jurisdictions among different actors. In Germany, mass communication or "distributed communication" (radio and TV) - as opposed to individual communication such as the telephone, etc. - falls under the jurisdiction of the federal states (the *Länder*). Individual communication, in contrast, is subject to central government control. Since from its early beginnings Bildschirmtext had been thought of as an electronic newspaper, the regulatory arena in Germany quickly widened to include the Länder. It is this arena extension which involved the media policy makers, who then proceeded to use their regulatory powers extensively. In the convention finally concluded between the federal government and the Länder, the issue of regulatory jurisdiction found a compromise solution, but it is quite likely that more extensive regulation of *Bildschirmtext* than might have resulted from a more centralized institutional structure has been the price.

5 Conclusion: Planned system development and the role of demand

The preceding comparison of videotex development in three countries has shown how the legal, political, and economic context shapes the strategies of major actors, who jointly determine the design of the evolving socio-technical system. In this concluding section we want to draw attention to some aspects of systems development considered as a collective decision process unfolding over time.

Videotex development in the three countries studied has dominantly been a top-down process: it emerged not "spontaneously", but was - in the three countries to a different degree - centrally planned.

This has a number of important implications. To begin with, large, spatially extended videotex systems do not seem to emerge where the state (or a national telecommunications monopoly) does not assume the role of system builder; this is shown by the lack of videotex development in the US[39]. This is, however, not a technical problem: Given the present state of technological development, videotex is not a "natural monopoly" and hence a collective good that *must* be centrally provided. If existing transmission networks can be used (leased, rented), videotex provision is not even prohibitively expensive. But from the very beginning there seemed to be no widespread demand among potential users for this particular service (or rather, bundle of information, transaction, and communication services). Some critics have even called videotex "a solution looking for a problem".[40] In any case, "market pull" has played only a minor role in this technological innovation process. This, however, does not mean that "technology push" can explain videotex development - the explanation rather has to be sought in the political field.

If the political or state actors did not initially *respond* to a perceived public demand, they still needed to *create* it for the new system to be viable. It is essentially the way in which the system has been set up and the way in which the demand has been created that explains the difference between France and the other countries. Whereas Germany and Britain used an incentive concept, France used a voluntaristic Trojan horse strategy, attempting to create an initial imperative need by introducing *Teletel* as a substitute for an earlier and essential product, the telephone directory. For this purpose France established a system which was fully integrated vertically and in which the French PTT controlled each component - from the terminal to the directory service. In Germany and Britain only the networks and databases were provided centrally - the growth of the terminal market and the information market has been left to market forces. So far, these forces have not been strong enough to overcome by themselves the critical threshold in the growth process. Certainly, in France the market of additional telematic services around the *annuaire électronique* is guided by the invisible hand, too, but on the basis of an already existing terminal park (beyond the threshold level) and a flexible and smoothly functioning computer network. In Britain and Germany the users still seem to show no vivid demand. But demand depends also on the quality of the services offered. The quality of services - hence indirectly the investment in services - depends in turn on the existing terminal park which defines the boundaries of this new market. Following the German and British introduction strategy, the users in these countries have to buy their terminals on a market where the costs are still high since the small volume and the large number of competitors prevent economies of scale. In Germany and in Britain there is, in fact, a double "chicken-egg dilemma" by which the service market *and* the terminal market are blocked and both blockages are closely related.

All three governments started from similar assumptions regarding the latent demands that might be stimulated by the new service. It is fascinating to observe how these "expectations of latent demand" were formed, partly disappointed, and revised. Initially, videotex was perceived primarily as a cheap and easy means of access to a large variety of useful information - information that could be highly specialized, detailed, and was always up to date. This functional image is reflected in the terminal configuration which was initially designed for accessing databanks with simple numeric keyboards. In addition, videotex was seen to provide small users with access to data-processing facilities - at a time when access to large computer centers was prohibitively expensive and difficult for small business users and non-computer specialists. Finally, it was thought that videotex might save time by making teleshopping, telebanking, and similar transactions possible. Thus, videotex appeared as a "cold" medium, an instrument of rationalization. This functional image reflects the logic of engineers, but also the logic of producers who are in the information and service market (e. g. newspapers, banks, travel agents, and mail order houses), and want to give "value for money" to their exacting clients.

Correspondingly, private households as well as professionals and firms were addressed as target groups because each of them was perceived as potentially having some of these information and transaction needs. Nobody seems to have doubted that professionals and small firms would quickly avail themselves of the new service. It appeared more difficult to open up the mass market of private households. Therefore the system builders in all three countries paid attention to this particular target group in their campaigns, field tests, and demonstrations of the new service. But only in France did the decision makers adopt a strategy based on a realistic assessment of the obstacles to a speedy diffusion in the private household sector. In Britain and Germany, on the other hand, we observe redefinitions of videotex as a medium for professional and business use as initial expectations of domestic utilization were disappointed.

The process of videotex development has not yet run its full course. Accordingly it is too early yet to tell whether existing differences among the three national videotex systems will persist or will rather be attenuated and disappear in the future. One reason for the latter to happen might be cross-national learning, where Britain and Germany could try to imitate the more successful French model. But at least in Britain, this is not what seems to be happening. The gradual shift of the German and British videotex systems toward professional applications is probably supporting a trend for videotex to lose its distinct identity. Initially seen as the prelude of the "home information society", videotex is today merging with conventional office information and communication systems. In the context of an ever growing number of competing services the distinct identity of *Prestel* is presently eroded and it becomes just one more brand among many offering comparable products. This process is largely the consequence of British deregulation and privatization policy, a road which France actually began to follow and Germany might eventually follow - even if with less determination.

An important factor facilitating the development in Britain is the growing diversification in the terminal field, where the spread of relatively cheap personal computers not only in offices, but also in the home makes recourse to the adapted TV set less and less necessary for using videotex. This holds equally for Germany and may soon even supersede not only the simple specialized videotex terminal but also the relatively "unintelligent" compact terminals as the French *Minitel* and the German *Bitel* or *Multitel*. More is at stake than the disappearance of a difference in the videotex terminals used in France on the one hand, and Britain and Germany on the other. The multifunctional home computer should facilitate access to many different information, transaction, and communication services (as it evidently does in Britain). It might in fact ease the growth of competing services so that in the end the bundle of functions that videotex is today might either be untied, or integrated into an even more comprehensive interactive telecommunications system.

Notes

1 This would be true for broadcast videotext. This text information service is offered and transmitted by TV stations and generally received on the (adapted) TV screen.
2 In view of the existing design alternatives, a general definition of videotex is rather difficult. One definition was elaborated within the "International Telegraph and Telephone Consultative Committee" (CCITT). According to this organization, a videotex system should have the following essential characteristics: "1) information is generally in an alphanumeric and/or pictorial form; 2) information is stored in a data base; 3) information is transmitted between the data base and users by telecommunication networks; 4) displayable information is presented on a suitably modified television receiver or other visual display device; 5) access is under the user's direct or indirect control; 6) the service provides facilities for users to create and modify information in the data bases; 8) the service provides data base management facilities which allow information providers to create, maintain and manage data bases and to manage closed user group facilities." CCITT, Telematic Services: Operations and Quality of Service (Recommendations F.160-F.350). Geneva: CCITT 1985, p. 88.
3 Tom Forester, High Tech Society: The Story of the Information Technology Revolution. Oxford: Blackwell, 1987, p. 126; Sam Fedida and Rex Malik, The Viewdata Revolution. London: Associated Business Press, 1979, p. 7: "Viewdata (...) will be one of the key systems of the 'silicon revolution', which in turn is one of the cornerstones of 'The Information Society'".
4 For some descriptions of viewdata see: James Martin, Viewdata and the Information Society. Englewood Cliffs/N.J.: Prentice-Hall, 1982. Margaret Bruce, British Telecom's Prestel. Material for Open University Course T362 (Design and Innovation). Milton Keynes: Open University Press, 1986. Rex Winsbury (ed.), Viewdata in Action: A Comparative Study of Prestel. London: McGraw-Hill, 1981.
5 For descriptions of Teletel see Claire Ancelin and Marie Marchand, Le Videotex: Contribution aux débats sur la télématique. Paris: Masson, 1984. La télématique grand public: rapports de la commission de la télématique au ministre des PTT.

Paris: La Documentation Française, 1986. Marie Marchand (ed.), Le paradis informationnels: Du minitel aux services de communication du futur. Paris: Masson, 1987. Marie Marchand, La grande aventure du minitel. Paris: Larousse, 1987. For a rather technical description of Teletel see Arun Ray-Barman, L'ère du videotex: Les outils de l'entreprise. Paris: Editests, 1985.

6 For a general description Dieter Lazak, Bildschirmtext: Technische Leistung und wirtschaftliche Anwendung neuer Kommunikationstechnik. München: CW-Publikationen, 1984.

7 The joint project, which began in the early part of 1986, is scheduled to last three years, and is being carried out by research teams based at the Max-Planck-Institut für Gesellschaftsforschung (MPI), the Centre Nationale de la Recherche Scientifique (CNRS) and the Science Policy Research Unit (SPRU). The teams are being financed by, respectively, the Deutsche Forschungsgemeinschaft and the Max-Planck-Gesellschaft, the French Administration, and the Leverhulme Trust. In France the work is carried out by Jean-Marie Charon and Thierry Vedel; in Britain by Ian Miles and Graham Thomas. We thank our colleagues for providing us with valuable information and research papers. The present paper profited from unpublished papers from and discussions with them: Jean-Marie Charon, Les acteurs du videotex français (1976-1986), Unpublished manuscript (ADI/Cristal:86/C03650), Paris, 1987. Graham Thomas, Prestel: Configuration, Regulation and Outline of Strategies, Unpublished working paper, Cologne, October 1986. Thierry Vedel, Le programme télématique français: l'apparition d'un nouveau medium, Unpublished working paper, March 1987.

8 Thomas Hughes, Networks of Power: Electrification in Western Society 1880-1930. Baltimore/Md., London: Johns Hopkins University Press, 1983.

9 A large part of this capacity is wasted because only a small part of information pages account for the overall use of the system. In PRESTEL only 100 pages account for about 25% of all information calls; 1,000 pages for 45% and 30,000 pages for 90%.

10 For Butler Cox cf. BTX-Aktuell (1987)7, p. 32-33.

11 The regional data bases have to store only a part of the total pages, and data transmission between regional centers is only necessary for pages which are seldom called. For a description of the IBM system, see: Ekhard Ording and Hans Hölsken, "Die neue BTX-Zentralentechnik: Das Konzept der IBM Deutschland GmbH und seine Realisierung", Fernmeldepraxis 12(1984), p. 480-488; see also Lazak 1984, op. cit.

12 For a detailed description of the German standard see Lazak 1984, op. cit.

13 British Telecom was privatized in 1984; at this time all relevant facets of the British Prestel had been decided.

14 Thierry Vedel (1987 op. cit.) writes about French regulation: "Depuis le 1er janvier 1986, l'offre de services télématiques sur le réseau de télécommunication est libre. La seule contrainte à laquelle est soumise une société prestataire de services télématiques est une déclaration préalable permettant d'identifier ses responsables légaux et la vocation du service. En outre une consultation télématique est considérée comme une communication privée et son contenu ne peut être censuré. Ce libéralisme - inspiré du modèle réglementaire de la presse écrite - a permis la création d'une multitude de sociétés de services télématiques ...".

15 See for example Eric Arnold, "Information Technology in the Home: The Failure of Prestel", in Niels Bjorn-Anderson, Michael Earl, Olav Holst and Enid Mumford (eds.), Information Society: For Richer, For Poorer. Amsterdam: North-Holland Publishing Company, 1982. For a comparative view see also: Eric Arnold and

Godefroy Dang-Nguyen, "Videotex: much noise about nothing?" in Margaret Sharp (ed.), Europe and the New Technologies. London: Frances Pinter, 1983.
16 Tom Forester 1987, op. cit.
17 See Luc Brunet, "Minitel: A Success Story" in W. Kaiser (ed.), Telematik (Telematica Kongreßband II). München: Fischer, 1986.
18 Cf. Lettre de Teletel, hors serie (1986)3; Listel: Repertoire des services Teletel, (January 1987)5.
19 Cf. Bureau et Informatique (April 1986)104.
20 From the use patterns of Teletel tariff system Teletel 1, which is used almost exclusively for professional services, one can conclude that Teletel has established itself in the professional world: from 226,000 hours of connection in Teletel 1 in January 1986 to 284,000 hours in December 1986 (cf. Lettres de Teletel 1986, op. cit.).
21 A market research study conducted at the end of 1985 concluded that 8,000 of the 30,000 users at that time were private. This figure, however, does not seem to be very reliable (cf. Btx Praxis (1986)1). A very recent survey indicates that only 20% of BTX terminals are privately used, 57% are professionally used and 23% are used for professional and private purposes. (BTX-Praxis (1988)5).
22 The figures are from the Prestel frame 65659a of July-September 1986.
23 This is undoubtedly supported by the tariff structure of Prestel, where business users have to pay more standing charges than private users.
24 According to a research study conducted by the Fraunhofer-Institut für Systemtechnik und Innovation, almost 80% of all service providers say that there would be no application which is economically profitable: ISI (Fraunhofer-Institut für Systemtechnik und Innovationsforschung), "Sozialhumane Auswirkungen von Bildschirmtext in ausgewählten Branchen". Bericht an das WIK. Karlsruhe, 1987.
25 Roy D. Bright, "Videotex - What Comes Next" in Diebold (ed.), Bildschirmtext-Kongreß 1982 - Proceedings. Frankfurt/M.: Diebold, 1982.
26 As Zysman put it: ".. the French are pursuing political goals in industrial development" and the French state is "accustomed to imposing its will on the marketplace." John Zysman, "Between the market and the state: dilemmas of French policy for the electronics industry", Research Policy 3(1975), pp. 312-336.
27 In the design of this strategy the report of Simon Nora and Alain Minc (L'informatisation de la société. Rapport à M. le Président de la République. Paris: Édition du Seuil, 1978) played an important role.
28 Lorenzi describes this basic idea as follows: "L'idée, héritée du rapport Nora-Minc, est simple. En électronique, il y a trois secteurs de base qui de plus en plus s'interpénètrent. Et il y a donc trois axes pour attaquer un même objectif: la société informationnelle de demain. La France est mal située sur deux d'entre eux: l'informatique (suprématie américaine) et les biens audio-visuels (suprématie japonaise); restent les télécommunications, c'est-à-dire les réseaux de transmission des informations." Jean-Hervé Lorenzi "Analyse d'un plan industriel au travers du concept de filière", Annales des Mines (January 1980), p. 80.
29 See ISI 1987, op. cit., p. 413 f.
30 See for example Jerome Aumente, New Electronic Pathways: Videotex, Teletext and Online Databases. Newbury Park/Calif.: Sage, 1987, p. 35.
31 For this argument see Bernard Corbineau, Espace politique et nouvelles technologies de l'information, Paper presented to the ECPR Joint Session of Workshops, Amsterdam, April 10-15, 1987, p. 3.
32 For the strategy of the UK TV industry see Colin Tipping, "Viewdata and the television industry" in Winsbury 1981, op. cit.

33 In Britain and in Germany, the consumer electronics industry seemed to be interested only in passive protection. For an active development policy the market seemed to be too insecure. The behavior of the British TV industry may also be explained by the fact that this industry funnelled its research and development funds in the development of the broadcast videotex TV terminals (Teletext) and Prestel consequently was given only subordinate priority.
34 Cf. the title of an article in Btx Praxis (1988)2, p. 41: "British Telecom schreibt Billig-Terminal aus: UK goes Minitel".
35 For a comparative view of the French industrial policy see Diana Green, "Promoting the Industries of the Future: The Search for an Industrial Strategy in Britain and France", Journal of Public Policy 1(1981)3, pp. 333-51. Diana Green, "Strategic Management and the State" in Kenneth Dyson and Stephen Wilks (eds.), Industrial Crisis: A Comparative Study of the State and Industry. London: Blackwell, 1983. Christian Stoffaës, "French Industrial Strategy in Sunrise Sectors" in Alexis Jacquemin (ed.), European Industry: Public Policy and Corporate Strategy. Calendron Press: Oxford, 1984.
36 See Zysman 1975, op. cit., p. 320. See also: Rex Malik, "The French keep to their grand strategie", Intermedia 11(1982)3, p. 12.
37 See Alan Cawson, The Teletext Initiative in Britain: The Anatomy of Successful Neo-Corporatist Policy-Making, Paper presented at the ECPR workshop on Meso-Corporatism, Amsterdam, April 1987.
38 For a comparative view of governmental programs in the communications domain, see Peter Humphreys, "Legitimating the Communications Revolution: Governments, Parties and Trade Unions in Britain, France and West Germany" in Kenneth Dyson and Peter Humphreys (eds.), The Politics of the Communications Revolution in Western Europe. London: Frank Cass, 1986, pp. 163-194.
39 For a recent overview on the U.S. videotex development see James Miller and Thierry Vedel, The Introduction of Videotex: The Role of the State in Canada, France and the U.S., Unpublished paper, 1987. See also John Tydeman et. al., Teletext and Videotex in the United States: Market Potential, Technology, Public Policy Issues. New York: McGraw-Hill, 1982.
40 See Forester 1987, op. cit., p. 126.